Early Praise for

EINSTEIN'S GOD

"Krista Tippett has a knack for finding thinkers who tackle deep and important questions in a sober but uninhibited fashion. The result is an exhilarating exploration of the meaning of it all."
—Robert Wright, author of *The Evolution of God*

"No one has a better ear for the most interesting facets of faith than Krista Tippett. And few topics lend themselves better to her nuanced interviews than the clash/collaboration/interplay of science and religion. If you want something beyond black-and-white culture war battles, you'll find these interviews powerfully stimulating."
—Steven Waldman, founder and editor in chief, Beliefnet

"In this sparkling book of interviews, Krista Tippett demonstrates that science and religion both benefit from a genuine dialogue. It doesn't matter if Tippett is talking about free will or the anatomy of the soul—she is always probing, measured, and illuminating. This book is a hopeful reminder that the intellectual conflicts we take for granted don't need to exist."
—Jonah Lehrer, author of *How We Decide*

continued . . .

＊

And acclaim for Krista Tippett's first book,

SPEAKING OF FAITH

"In a day where religion—or, rather, arguments over religion—
divide us into ever more entrenched and frustrated camps, Krista
Tippett is exactly the measured, balanced commentator we need.
Her intelligence is like a salve for all thinking people who have felt
wounded or marginalized by the God Wars."

—Elizabeth Gilbert, author of *Eat, Pray, Love*

"We need to take religion seriously. We need to give voice in
our national life to religious and nonreligious people alike who
understand, as Tippett does, that the crooked line dividing good
from evil runs through each of us . . . We should be grateful to
Tippett not only for amplifying voices of sane faith but also for
modeling in herself a public theology that manages to derive faith
out of doubt and hope out of paradox."

—Stephen Prothero, author of *Religious Literacy*

"At a time when professional contrarians like Sam Harris and
Christopher Hitchens take the meaning and mystery out of religion,
Krista Tippett is a welcome voice of literate faith."

—*The Dallas Morning News*

"*Speaking of Faith* is a chronicle of ideas . . . It is also a memoir with
a narrative thread that makes this book as compelling as any novel.
In her book, as on the radio, Tippett is an intelligent observer,
sophisticated and compassionate. Reading Tippett is like attending a
dinner party with some of the most interesting minds in America and
standing at the elbow of the hostess as she introduces each friend."

—Catholic Online

EINSTEIN'S GOD

CONVERSATIONS ABOUT SCIENCE
AND THE HUMAN SPIRIT

KRISTA TIPPETT

PENGUIN BOOKS

PENGUIN BOOKS

Published by the Penguin Group
Penguin Group (USA) Inc., 375 Hudson Street, New York, New York 10014, U.S.A.
Penguin Group (Canada), 90 Eglinton Avenue East, Suite 700, Toronto, Ontario, Canada M4P 2Y3
(a division of Pearson Penguin Canada Inc.) · Penguin Books Ltd, 80 Strand, London WC2R ORL,
England · Penguin Ireland, 25 St Stephen's Green, Dublin 2, Ireland (a division of Penguin Books
Ltd) · Penguin Group (Australia), 250 Camberwell Road, Camberwell, Victoria 3124, Australia
(a division of Pearson Australia Group Pty Ltd) · Penguin Books India Pvt Ltd, 11 Community
Centre, Panchsheel Park, New Delhi – 110 017, India · Penguin Group (NZ), 67 Apollo Drive,
Rosedale, North Shore 0632, New Zealand (a division of Pearson New Zealand Ltd) · Penguin
Books (South Africa) (Pty) Ltd, 24 Sturdee Avenue, Rosebank, Johannesburg 2196, South Africa

Penguin Books Ltd, Registered Offices:
80 Strand, London WC2R ORL, England

First published in Penguin Books 2010

5   7   9   10   8   6

This book contains excerpts from interviews that were first broadcast on Speaking of Faith®,
copyright © American Public Media. Used by special permission.

"Questo Muro" and "Heart Work" by Anita Barrows. Used by permission of the author.
Excerpts from A Madman Dreams of Turing Machines by Janna Levin. Copyright © 2006 by Janna
Levin. Used by permission of Alfred A. Knopf, a division of Random House, Inc.
"Dich wundert nicht. . ./ You are not surprised. . ." from Rilke's Book of Hours:
Love Poems to God by Rainer Maria Rilke, translated by Anita Barrows and Joanna Macy.
Copyright © Anita Barrows and Joanna Macy, 1996. Used by permission of
Riverhead Books, a member of Penguin Group (USA) Inc.

LIBRARY OF CONGRESS CATALOGING IN PUBLICATION DATA
Tippett, Krista.
Einstein's God : conversations about science and the human spirit / Krista Tippett.
p. cm.
ISBN 978-0-14-311677-6
1. Religion and science.   2. Scientists—Interviews.   I. Title.
BL240.3.T57 2010
215—dc22   2009043521

Printed in the United States of America
Set in Perpetua • Designed by Elke Sigal

For Trent Gilliss, Mitch Hanley, Colleen Scheck,
and Kate Moos Kate Moos Kate Moos—fantastic colleagues
and beloved friends who built this thing with me

CONTENTS

✳

Introduction · *1*

1

THE HUMAN LEGACY OF A GREAT MIND AND A WISE MAN
Einstein's God
with FREEMAN DYSON *and* PAUL DAVIES · *15*

2

THE SPIRIT AS AN EMERGENT LIFE FORCE
The Biology of the Spirit
with SHERWIN NULAND · *41*

3

DISCOVERING THE GLOBALIZATION OF MEDICINE
Heart and Soul
with MEHMET OZ · *65*

4

CREATION AS AN UNFOLDING REALITY
Evolution and Wonder
with JAMES MOORE · *95*

5

CONTENT WITH THE LIMITS OF RELIGION AND SCIENCE
The Heart's Reason
with V. V. RAMAN · *121*

6

THE WORLD FEELS MORE SPACIOUS
Mathematics, Purpose, and Truth

with JANNA LEVIN · *143*

7

SCIENCE THAT LIBERATES US FROM REDUCTIVE ANALYSES
Getting Revenge and Forgiveness

with MICHAEL McCULLOUGH · *171*

8

KNOWING HOW TO HEAL OURSELVES
Stress and the Balance Within

with ESTHER STERNBERG · *197*

9

THE NATURE OF HUMAN VITALITY
The Soul in Depression

with ANDREW SOLOMON, PARKER PALMER,
and ANITA BARROWS · *223*

10

ON THE COMPLEMENTARY NATURE OF SCIENCE AND RELIGION
Quarks and Creation

with JOHN POLKINGHORNE · *251*

About the Interviewees · *281*

Acknowledgments · *285*

EINSTEIN'S GOD

INTRODUCTION

✳

*T*he science-religion "debate" is unwinnable, and it has led us astray. To insist that science and religion speak the same language, or draw the same conclusions, is to miss the point of both of these pursuits of cohesive knowledge and underlying truth. To create a competition between them, in terms of relevance or rightness, is self defeating. Both science and religion are set to animate the twenty-first century with new vigor. This will happen whether their practitioners are in dialogue or not. But the dialogue that is possible—and that has developed organically, below the journalistic and political radar—is mutually illuminating and lush with promise. This book is a conversational introduction to an interplay between scientific and religious questions—not as argued, but as lived—that I began to discover a decade ago.

At that time, in the late 1990s, I started a media experiment that eventually became a weekly public radio program about religion, ethics, and questions of meaning, *Speaking of Faith*. I wanted to explore the intellectual and spiritual content of this part of life we call "religious" and "spiritual" and all the complexity with which it finds expression. Since the passing of Niebuhr and Heschel, of Tillich and King, we had lost a robust vocabulary for spiritual ethics and theological thinking in American public life. In polite, erudite, public-radio-loving circles, religion had become something, as the sociologist Peter Berger quips, "that was done in private between consenting adults."

I came to adulthood in such a milieu and never questioned its rightness. I went to Brown, studied Ostpolitik in Bonn, landed in divided Berlin as the *New York Times* stringer, and spent most of the eighties there, most of my twenties, as a journalist and then a diplomatic appointee. Politics on that cold war fault line was morally as well as strategically thrilling. Spiritually I was agnostic, I suppose, though I'm not sure I gave religion enough thought in those years to claim the label.

Yet I had grown up in the intellectual and spiritual domain of the Jerry Falwells and Pat Robertsons of the world. Like them, my grandfather was a preacher of hellfire and brimstone. At the same time, though the product of a second grade education, he had a large, unexcavated mind that frightened him, I think, but fascinated me—a sharp wit, a searching attentiveness, a mysterious ability to perform mathematical feats in his head. People like my grandfather were badly represented by Jerry Falwell and Pat Robertson and the journalists who gave them powerful platforms in the eighties and nineties. Later, perhaps understandably, people like him became the object of erudite parody, straw men easily blown down by prophets of reason. His kind of religiosity was small-minded at best, delusional at worst, and, most damnably, the enemy of science.

The mundane truth is this: my grandfather did not know enough about science to be against it. I summon his memory by way of tracing, for myself, why I've found my conversations with scientists to be so profoundly sustaining. It is not just that they are intellectually and spiritually evocative beyond compare. Cumulatively they dispel the myth of the clash of civilizations between science and religion, indeed between spirit and reason, that we've accepted as the backdrop for so many tensions of the modern West.

In the beginning, I sought out people with an overt passion to reconcile science and religion in their discipline and in their person. Sir John Polkinghorne is one of the most prominent of these globally—a Cambridge quantum physicist who also became a Cambridge theologian in midlife and has written eloquently about finding both science and religion necessary to interpret the "rich, varied and surprising way the world actually is." I found his approach revelatory as I was cautiously finding my own way back to religion after Berlin. As a physicist, Polkinghorne sees a universe that is "supple" and "subtle"—a mix of determinism and of freedom—and this informs his imagination about the nature of God, what happens when we die, and what happens when he prays.

But as the years progressed I've been equally intrigued, and driven to new places in my own thinking, by scientists like the theoretical physicist and novelist Janna Levin. She is exploring the shape and finitude of the universe. She is fascinated by mathematical insights into how we can know what is real and true and how free we really might be. She is not a religious person in any sense, but her scientific inquiry is philosophically and spiritually evocative, rich in the raw materials of theology.

Albert Einstein was more like Janna Levin than John Polkinghorne. His famous quip that "God does not play dice with the universe" is often wrongly imagined as a statement of faith, when in fact it was a clever barb tossed in a strictly scientific argument. Focusing as he did on the evolution of stars and galaxies and on intangible substances of light, time, and gravity, Einstein seemed to present little to offend religion. But as much as or more than Darwin's natural laws of evolution, Einstein's laws of physics could not tolerate a meddling divine hand.

Einstein approached science itself with a religious awe, as the physicist Freeman Dyson tells us. Yet as a young colleague of Einstein at Princeton, Dyson saw him become more philosophical as he grew older, leaving behind a rich body of reflection on the "mind" and "superior spirit" behind the cosmos. And as the astrophysicist Paul Davies describes in these pages, modern imaginations have yet to catch up to the potential spiritual implications of the way Einstein reframed our understanding of space and time. Einstein's dismissal of a "personal God" might have struck some in his time as heretical, but his self-described "cosmic religious sense" is intriguingly resonant with twenty-first-century sensibilities. There has simply been too little space in our public life up to now to hear such echoes.

Here, as in so many other realms of life, a wider lens of perspective can make all the difference. For example, it is important to see—though this basic fact is rarely invoked alongside global characterizations of the "religion versus science" scenario—that only in Christianity were defining battle lines drawn after the Enlightenment between the forms of knowledge that religion and science pursue. Those battle lines galvanize a few of the traditions of Christianity and others inconsistently or not at all. The first presiding bishop of the U.S. Episcopal Church to be elected in this century is a marine biologist by training. The scientist who presided over the Human Genome Project that first mapped human DNA is an evangelical Christian.

Antiscience perspectives are even more marginal in the sweep of the world's great religious and spiritual traditions. There are few strident Jewish voices in the science-centered "moral values" debates of American culture of recent memory, from abortion to stem-cell research. And there are theological, not merely cultural, reasons for this. Religious virtues of "justice" and "healing" weigh heavily in discerning the manifold implications of "the sanctity of life." Islamic theology similarly offers a distinctive approach to issues such as evolution and the moral status of the fetus, hence the lack of famously strident Muslim antiscience voices. The physicist V. V. Raman describes in these pages how Hinduism's overarching regard for beauty and the arts has helped it avoid a point-counterpoint between the different forms of knowledge that science and religion convey. Hinduism's offspring, Buddhism, is in a class of its own. Einstein liked to imagine Buddhism as the religion of the future, capable of embracing the best of scientific and spiritual approaches to life. In recent decades, Buddhist spiritual technologies of mindfulness and meditation have presented themselves with transforming effect in Western lives and Western medicine.

As we bring this debate closer to the ground, in fact, and expose it to the plain light of the everyday, the suggestion that science and religion are incompatible makes no sense at all. In the vast middle of modern Western culture, scientific and religious insights coexist and intertwine for the most part peaceably. We encounter and respond to the fruits of science in our doctors' offices; through experiences of birth, illness, and death; in the ever-evolving technology at the center of ordinary life. Opinion polls promote hyperbole and false dichotomies. Ask Americans to choose between God and Darwin and they'll opt for God. But generations of Christian Americans have also grown up learning about evolution in scientific textbooks and about a God behind creation in the biblical book of Genesis—and intuitively reconciling them, instinctively imagining that both might simultaneously be true.

As both John Polkinghorne and the Darwin biographer James

Moore describe in these pages, Genesis is in fact a compelling example of how treating sacred text seriously, reading it respectfully on its own terms, is the surest, strongest antidote to our polarized religio-cultural debates. This is a text infused with purpose, but that purpose was not to narrow our pursuit of understanding the natural world. For centuries, until the medieval period and the Reformation, great Christian theologians knew this and honored it. To treat Genesis as a commentary on science is to ignore its cogency as text and teaching, just as to read a poem as prose is to miss the point. It is more complicated than that, but it is also that simple.

And just as a more three-dimensional approach to the Bible can provide new starting points for an old conversation, so can a more three-dimensional look at the history of science. Even when they struggled against bitter religious resistance to their ideas, the likes of Copernicus, Galileo, Kepler, and Newton believed that their discoveries would and should widen human comprehension of the nature of God. The more we could understand about the world around us in all its intricacy, their reasoning went, the better we would understand the mind of its maker.

Charles Darwin belonged to that lineage. *The Origin of Species* was not the first text to break from religion, as our cultural narrative has come to assume. It was the last classic scientific text to engage theology directly. James Moore lays this out forcefully. And for the religious scientists in these pages, no intellectual compromise is needed to embrace evolution as ingenious—to understand creation as an ongoing, inborn capacity of a world endowed with independence rather than as the one-act invention of a puppet-master God. James Moore also makes the compelling suggestion that in documenting the freedom of the world to define its own fruitfulness in and through chaos and struggle,

Darwin liberated humanity from belief in a God who preordained every cancerous cell and shifting tectonic plate, every social and physical injustice. Even the creationists of our time have been liberated—in part by Darwin—from belief in this kind of God.

Images from the world of science enliven my understanding of God, and of religion. The wildly imaginative discipline of physics alone, as evident in these pages, is rife with starters. Contemporary physics revolves around objects, premises—quarks, for example, and strings—that no one has ever seen or expects to "see"; but worlds of passion and discovery and progress thrive on them, because the idea of them gives intelligibility to the whole of what can be measured, experienced, and observed. A scientific puzzle that Einstein chewed on, the question of whether light is a particle or a wave, was resolved by a teacher of John Polkinghorne, Paul Dirac, with the unexpected, seemingly illogical conclusion that it is both. And here's the key that made that discovery possible: how we ask our questions affects the answers we arrive at. Light appears as a wave if you ask it "a wavelike question" and it appears as a particle if you ask it "a particle-like question." This is a template for understanding how contradictory explanations of reality can simultaneously be true.

And it's not so much true, as our cultural debates presume, that science and religion reach contradictory answers to the same particular questions of human life. Far more often, they simply ask different kinds of questions altogether, probing and illuminating in ways neither could alone. The biologist Carl Feit of Yeshiva University helped me understand this first, in the earliest days of my life of conversation, describing his clinical pursuit of cancer and his religious study of the Talmud as "dual intellectual quests." Science asks penetrating questions of "how," he explains, yet,

The physical universe doesn't come beset with values. It's kind of neutral in the sense that it can be used for good and for bad. From the scientific perspective, everything that we can discover we should discover. The problem comes up, what do you do with something once you've discovered it? That was the moral dilemma faced by the scientists after World War II . . . when they realized that they were working on exploring and exploiting the potential energy that's present in an atom. That's a two-edged sword that could be used to destroy humanity. But it also can be used to cure cancer.

V. V. Raman's mother tongue of Tamil linguistically distinguishes between the word "why" as a causative question—the way science approaches a problem—and "why" as an investigation of purpose—the way religion might approach the same problem, with very different results.

The religious impulse is animated at its core by questions of purpose: What does it mean to be human? Where do we come from? Where are we going? How to love? What matters in a life? What matters in a death? How to be of service to one another and to the world? As the immunologist Esther Sternberg and the cardiologist Mehmet Oz realized at turning points in their professional lives, the scientific core of Western medicine cannot resolve or even really address the vulnerability of human life, the inevitability of death, our ordinary and persistent struggles for meaning in between. But traditions and practices of faith accompany these, face them, bless them. The anthropology of faith—its insistence that critical aspects of life are unquantifiable, mysterious, and blessedly imperfect—puts it squarely in the camp of reality if not of logic.

Clear-eyed wisdom about the human condition is not religion's most famous attribute in our time, but it is there in the DNA of the theology I love and pursue, and it could help reframe faith's encounter with science. Consider the perfect opening line of Reinhold Niebuhr's twentieth-century theological classic *The Nature and Destiny of Man*: "Man has always been his own most vexing problem." I hear this as a succinct diagnosis of Einstein's dismayed observation, as chemists and physicists became eager purveyors of mid-twentieth-century weaponry, that technology in his generation was like a razor blade in the hands of a three-year-old. One cannot lead an examined life without noticing that all of our grandest objectives—political, economic, and scientific—are inevitably complicated by the inner drama of the human condition. In this spirit, Einstein came to understand his contemporary, Mahatma Gandhi, and other figures such as Jesus, Moses, St. Francis of Assisi, and Buddha, as "spiritual geniuses"—"geniuses in the art of living . . . more necessary to the sustenance of global human dignity, security and joy than the discovers of objective knowledge."

I wonder if Einstein would be as fascinated as I am at how science in our age is yielding measurable insights into the tools of the trade of the spiritual geniuses of the ages. He might rather be shooting down string theory, or still pursuing his unified Theory of Everything, a phrase ripe for theological mulling if ever there was one. But my work moving forwards is also galvanized by my discovery, laid out in these pages, that on frontiers unfolding in our age, science is presenting whole new realms of challenge and promise for religion's self-understanding and its place in the world.

To reiterate: this is not happening because scientists are religious, or (God forbid, Richard Dawkins might say) setting out to help religion. It is happening organically as science yields ever more intricate tools to study humanity—pursuing its own versions of the animating question of what it means to be human. The wise physician and author Sherwin Nuland is Jewish, though not religiously devout, but he finds in the church father St. Augustine apt words about the reverence for the human physical experience that drives his work:

> *Men go forth to wonder at the heights of*
> *mountains, the huge waves of the sea,*
> *the broad flow of the rivers, the vast*
> *compass of the ocean, the courses of the*
> *stars: and they pass by themselves*
> *without wondering.*

And with the advent of scientific tools to study what Nuland calls the most remarkable of all things in nature, "the three-pound human brain," scientific modes of understanding are trained even on those aspects of human nature that religion cultivates at its best. So the clinical psychologist Michael McCullough can document how violent acts of revenge are both normal and purposeful, rooted in the physiology of humanity as much as in its history. But so, he and his colleagues are learning, is an instinct to forgive. This kind of research helps us define what we can't deny or control in terms of hardwired impulses towards violence. At the same time, it helps us rethink the control we do have to create conditions that will nurture or trigger empathy and forgiveness over violence or revenge. Key to that are the perceptions we hold of others—whether we find it in ourselves to see their "value" for us, to understand their well-being as linked to ours.

This science presents a pointed call to thinkers and leaders in the great religious traditions, but not to dust off obvious teachings on forgiveness and compassion. It is an illuminating revelation that these will "work" only if the traditions mine their equally ancient and powerful—but relatively neglected—repositories of practical resources to help us see and internalize the value of "the other"—whether enemy or friend, neighbor or stranger. This suggestion of using science as a tool for promoting lived religious virtue might strike some as cynical. But here again I'd insist that religion at its best is clear-eyed and reality-based. The most vivid saint in living memory, Mother Teresa, steeped herself in human death and decay and cared right there. Einstein developed a stronger sense of his Jewish identity as he moved through life, increasingly valuing Judaism's practical moral core, its fixation less on transcendence than on "life as we live it and can grasp it." Science can take a notion like altruism out of the realm of idealism—offering us a more sophisticated view of it than either religion or evolutionary biology have proposed heretofore. But science cannot mobilize human consciousness and human passion. We need simultaneous resources of story, ritual, relationship, and service that spiritual traditions have the capacity to nurture at their core. Our common life needs the moral vocabulary and practices that spiritual traditions have sustained across centuries and generations—of healing and repair, of repentance and reconciliation, of mindfulness and hospitality—as much as it needs sophisticated vocabulary for political, economic, and military endeavor.

This realization is coming full circle now in the realm of medicine. Where a scientific emphasis on what can be measured once took humanity away from a seriousness about spiritual and emotional aspects of human vitality, science itself is now bringing us back. In a handful of years, centers for spirituality and healing have emerged at major medical schools and hospitals across the United States. We can measure the effects of something like prayer and meditation on the brain and body.

We know that what we call feelings—both physical and

emotional—are caused by traceable biochemical connections. Working at this juncture of physiology and feelings, health and emotion, Esther Sternberg can provide a concrete understanding of the stress response to illuminate questions that religion has raised up to now but medicine had left hanging: How does stress make us sick? Conversely, why might places of peace, prayer, meditation, rest, music, and friendship help us to live well?

This kind of whole-mind, whole-body exploration of what it means to be human also opens up new ways to address the dark side of human experience. Depression is a widespread malady of our age, and an excruciating demonstration of mind, body, and spirit enmeshed. For centuries, Western culture defined and addressed depression in largely spiritual terms. In recent years, as we've learned about genes, hormones, and neurotransmitters, we came to discuss it overwhelmingly in biological, pharmacological terms. In three conversations exploring the notion of "the soul in depression" with the journalist and chronicler of depression Andrew Solomon, the Quaker author Parker Palmer, and the Buddhist psychologist and poet Anita Barrows, I find a kind of antidote and balance. All of them have been saved by medication in their own struggles with depression. They have also come to a reverence, from very different spiritual vantage points, for the paradoxical new awareness of the "soul" that they gained by way of its seeming absence in the depths of depression. Beyond those depths, and with the wisdom of years, all of them trace a treacherous but ultimately hopeful line between the illness of depression and the darkness that is a part of human vitality and that we can embrace. This conversation needs the insights of both science and spirituality to evolve.

One of my favorite twentieth-century theologians, Dietrich Bonhoeffer, was a contemporary of Albert Einstein. He thrilled to his readings of the physics of his day from the Nazi prison where he eventually died. He decried a stunted religious imagination that would consign God to the borders where scientific knowledge gives out. And after these years of conversation with scientists, a sentence that struck me from my earliest readings of Bonhoeffer comes back with new connotations: "I worry that Christians who have only one foot on earth can also only have one foot in heaven."

Having two feet on earth in our time means knowing about black holes and brain chemistry; it means pondering whether the universe is infinite or finite and what the matter in "dark matter" might be. My conversations with scientists leave me with an exhilarating sense of the immediacy and vastness of both reality and mystery, of the importance of asking seemingly unanswerable questions, and of the "rationality" of insisting on a world in which ethics, theology, and "spiritual genius" claim their place alongside and in collaboration with the wondrous capacities of science. To the faithful I say this: if God is God, we cannot be afraid of what we can learn with the remarkable three-pound brain. I offer this book to all—religious and nonreligious, theologians, scientists, and people of all walks of life in between—who want to engage our kindred capacities to think and to live together more richly than our debates would ever suggest is possible.

1

✳

The Human Legacy of a Great Mind and a Wise Man

"EINSTEIN'S GOD"

*A*lbert Einstein's famous equation, $E = mc^2$, remains difficult for me to grasp fully. But I feel I have come to understand something of the man—his expansive spirit, his relentless curiosity, and his reverence for the beauty and order of nature and thought. I was daunted as I began, but delving into Einstein was a delight.

And there is a logic of sorts to that, as humor was an aspect of Einstein's genius. Freeman Dyson suggests that his ability to make light and to laugh, even at himself, was one key to the magnitude of his scientific accomplishment. Science is often about failure. Einstein himself proposed that he made so many discoveries because he was not afraid to be proven wrong, repeatedly, on his way to all of them. But Einstein also employed humor to philosophical and ethical effect, weighing in trenchantly on mankind's foibles.

Einstein held a deep and nuanced, if not a traditional, faith. I did not assume this at the outset. I've always been suspicious of the way Einstein's famous line, "God does not play dice with the universe," gets quoted for vastly different purposes. I wanted to understand what Einstein meant as a physicist when he said that. As it turns out, that particular quip had more to do with physics than with God, as Freeman Dyson and Paul Davies illuminate.

Einstein did, however, leave behind a rich body of reflection on the "mind" and the "superior spirit" behind the cosmos that has never made its way into popular consciousness. He didn't believe in a personal God who would interfere with the laws of physics. But he was fascinated with the ingenuity of those laws and expressed awe at the very fact of their existence. Throughout his life, he thrilled to all he could not yet understand. He was more than content with what he called a "cosmic religious sense"— animated by "inklings" and "wondering," rather than by answers and conclusions. Here is a passage that comes close, I think, to a concise description by Einstein of his quintessential "faith":

> A knowledge of the existence of something we cannot penetrate, of the manifestations of the profoundest reason and the most radiant beauty—it is this knowledge and this emotion that constitute the truly religious attitude; in this sense, and in this alone, I am a deeply religious man. I cannot conceive of a God who rewards and punishes his creatures, or has a will of the type of which we are conscious in ourselves . . . Enough for me the mystery of the eternity of life, and the inkling of the marvelous structure of reality, together with the single-hearted endeavor to comprehend a portion, be it ever so tiny, of the reason that manifests itself in nature.

With Paul Davies, I was able to pursue how Einstein changed our view of space and especially time, a subject that has always intrigued me. Before Einstein, as Davies describes it, human beings thought of space and time as fixed and immutable, the backdrop to the great show of life. But we now know they are elastic and intertwined, part of the show themselves. Einstein described our perception of time as an arrow—traversing linear and compartmentalized past, present, and future—as a "stubbornly persistent illusion." Such language is evocative from a religious standpoint. As Davies discusses, it echoes insights that run throughout Eastern and Western religions and ancient indigenous cultures. Davies finds an affinity between Einstein's view of time and the religious notion of a reality "beyond time," and of "the eternal." And because he speaks as a person conversant in current advancements of Einstein's science—cosmology and the Big Bang, black holes, even the search for life beyond this galaxy—his insights carry for me a special weight of authority and, yes, wonder.

I came across many wise and touching pieces of writing by the spiritual Einstein while preparing for these conversations. Einstein was a passionate letter writer. He wrote to fellow scientists, friends, and strangers. He loved responding to the letters of schoolchildren. One of his correspondents for a time was Queen Elisabeth of Belgium. He had struck up a warm friendship with her and her husband, King Albert, just before World War II. In one tragic season in the midst of already tumultuous political times, her husband died suddenly, as did her daughter-in-law. Einstein wrote to her:

Mrs. Barjansky wrote to me how gravely living in itself causes you suffering and how numbed you are by the indescribably painful blows that have befallen you.

And yet we should not grieve for those who have gone from us in the primes of their lives after happy and fruitful years of activity, and who have been privileged to accomplish in full measure their task in life.

Something there is that can refresh and revivify older people: joy in the activities of the younger generation—a joy, to be sure, that is clouded by dark forebodings in these unsettled times. And yet, as always, the springtime sun brings forth new life, and we may rejoice because of this new life and contribute to its unfolding; and Mozart remains as beautiful and tender as he always was and always will be. There is, after all, something eternal that lies beyond the hand of fate and of all human delusions. And such eternals lie closer to an older person than to a younger one oscillating between fear and hope. For us, there remains the privilege of experiencing beauty and truth in their purest forms.

I emerged from these discussions with a new sense of Albert Einstein—not just as a great mind, but as a wise man. He was fully human and flawed, certainly in his intimate relationships. But he was undeniably an original, and not just as a scientist. If past, present, and future are an illusion, as he said, none of us ever really disappear. We all leave our imprint on what is now. I have a profound sense of Einstein's imprint, and it comforts me. I suspect that if he heard he was the subject of a program called *Speaking of Faith* more than fifty years after his death, he would make a funny, kindly, self-deprecating joke. But if he could listen with twenty-first-century ears, he might be intrigued by how his generous, questioning, "cosmic" religious sense is deeply kindred with the religious and spiritual yearnings of our age.

Einstein's God

KRISTA TIPPETT, host
FREEMAN DYSON, theoretical physicist and author
PAUL DAVIES, astrophysicist and author

✳

In 1905, a twenty-six-year-old examiner in the Swiss patent office in Bern made a series of discoveries that altered the course of modern science. Most famously, Albert Einstein proposed the theory of special relativity, which changed the way we think about space, time, and matter. The theory is best known by a single elegant equation: $E = mc^2$. Ten years later he took that a step further by accounting for the effects of gravity in his theory of general relativity. Though most of us can't grasp the full sense of general relativity, scientists agree that it describes the fabric of the universe we inhabit and that without Albert Einstein we still might not know it.

One of my guests, the astrobiologist Paul Davies, offers this analogy: "Until Einstein, people thought of time and space as fixed, unchanging, and absolute, the backdrop to the great show of life. Einstein revealed that time and space themselves are elastic and mutable, that they exist in relationship with unfolding life. They are part of the show themselves. Time, space, matter, gravity, and light are all intertwined. They curve and collapse and change in response to each other. Such insights gave rise to the grand ideas that occupy physicists and cosmologists today: the Big Bang, black holes, quantum mechanics."

Albert Einstein often attributed his genius to the fact that he was a late bloomer as a child. In consequence, he proposed, he remained enthralled into adulthood with elemental features of existence that most of us take for granted. Here's a passage from Albert Einstein's autobiographical notes published in 1949:

Why do we come, sometimes spontaneously, to wonder about something? I think that wondering to one's self occurs when an experience conflicts with our fixed ways of seeing the world. I had one such experience of wondering when I was a child of four or five and my father showed me a compass. This needle behaved in such a determined way and did not fit into the usual explanation of how the world works. That is that you must touch something to move it. I still remember now, or I believe that I remember, that this experience made a deep and lasting impression on me. There must be something deeply hidden behind everything.

After seeing that compass, Einstein became mesmerized in turn by light and gravity. He spent his life seeking to comprehend the order "deeply hidden behind everything" and to describe it mathematically. Einstein often spoke of this as his longing to understand what God was thinking.

When my first guest, Freeman Dyson, was born in England in 1924, Albert Einstein was at the height of his fame. As a young boy, Dyson yearned to speak Einstein's language of mathematics. He went on to become an eminent theoretical physicist at the Institute for Advanced Study in Princeton, where Einstein spent the last two decades of his life.

TIPPETT: Let's talk about the way Einstein used the word "God." He did seem to make frequent references to "the Lord." And he also said that what drove him all his life, what drove him as a scientist, was understanding if God had to make the world this way.

DYSON: Yes. Well, certainly it was not the kind of personal God that many people believe in. And he said that very explicitly, that he did not believe in a personal God who was interested in human affairs. He did believe in nature as some sort of universal spirit, or I suppose you might say "world soul," or some kind of universal mind, which ruled the universe and which was far beyond our comprehension. That's what he called "God" or "the Lord." He was not a practicing Jew, but he certainly knew that Jewish literature, and "the Lord" is a phrase that's used in the Bible, in the Old Testament.

TIPPETT: There's a kind of reverence in that term, isn't there, implicitly?

DYSON: Yes.

TIPPETT: You have written of yourself that you are a practicing Christian, but not a believing Christian. And it seems to me that Einstein might well have made the same statement about himself as a Jew.

DYSON: Well, he wasn't really a practicing Jew in that he didn't observe the Sabbath. But still, it was certainly true that he was a sort of a cultural Jew, but not a believing Jew.

TIPPETT: I'm quite intrigued by how he seemed to have developed a real reverence for Judaism, I guess, later in his life. That he saw it as a moral attitude in life and to life, not a transcendental religion. He wrote, "It is concerned with life as we live it and can, up to a point, grasp it, and nothing else." It seemed to him to be compatible with his faith, as you described it, as a scientist.

DYSON: Oh, yes. He took a very solemn view of science. And science was, to him, a religion. He said that quite explicitly. Of course, in later life he became much more philosophical than he was as a young man. But in later life, he said explicitly that anybody who does not approach science with religious awe is not a true scientist.

TIPPETT: When you say that you're a practicing Christian, but not a believing Christian, aren't you also saying that you don't need or even desire to pin down a theology? That you, as a scientist—and I think that Einstein was like you in this respect—that you are accustomed to and even thrilled by what you can't yet know or haven't yet discovered?

DYSON: Absolutely. The world is full of mysteries, and I love mysteries. Of course, science is full of mysteries. Every time we discover something, we find two more questions to ask, and so there's no end of mysteries in science. That's what it's all about. And the same's true of religion.

In an address at a conference on science, philosophy, and religion in 1941, Albert Einstein declared that science can be created only by those who aspire towards truth and understanding. He famously concluded: "Science without religion is lame. Religion without science is blind." Einstein understood science and religion to be separate realms, but joined by kindred impulses.

Most often he stressed how both realms acknowledge and honor the human sense of mystery, as in this passage from his autobiography *The World As I See It*, published in 1956:

> The fairest thing we can experience is the mysterious. It is the fundamental emotion which stands at the cradle of true art and true science. He who knows it not and can no longer wonder, no longer feel amazement, is as good as dead. A snuffed-out candle. It was the experience of mystery, even if mixed with fear, that engendered religion. A knowledge of the existence of something we cannot penetrate, of the manifestations of the profoundest reason and the most radiant beauty. It is this knowledge and this emotion that constitute the truly religious attitude. In this sense, and in this alone, I am a deeply religious man. I cannot conceive of a God who rewards and punishes his creatures, or has a will of the type of which we are conscious in ourselves. Enough for me, the mystery of the eternity of life and the inkling of the marvelous structure of reality, together with the single-hearted endeavor to comprehend a portion, be it ever so tiny, of the reason that manifests itself in nature.

In his greatest discoveries, Einstein focused on the laws that govern the largest dimensions and energies of physics—"the mountaintops," as Freeman Dyson puts it. But Einstein's work also opened physics to the study of the smallest quantum particles. And during Einstein's lifetime, quantum physicists such as Niels Bohr and Werner Heisenberg proceeded to find randomness and unpredictability in that sphere. In ordinary space, we throw a ball into the air and it comes back down. But at the atomic level, Heisenberg proclaimed, "anything could happen. Atoms veer off in wholly unpredictable, illogical directions, seemingly of their own will."

Einstein found this idea unacceptable. He drew the closest thing he had to a theology from his reverence for the writings of the seventeenth-century Dutch philosopher Baruch Spinoza. Spinoza described God's superior intelligence manifest in the determined harmonious order of nature. And Einstein spoke his most famous sentence about God as he disputed the disorderly universe of quantum physics. He said repeatedly, "I do not believe that God plays dice with the universe."

❋

DYSON: He had this religious faith, I would say, in the power of nature, and he saw nature as something causal so that, in some way, it was predetermined from the beginning of time how it was going to go on. And that is not the way we see things happening today.

TIPPETT: It's said that Einstein said to Niels Bohr, "God does not play dice with the universe," and Bohr responded, "Who is Einstein to tell the Lord what to do?"

DYSON: Yes. And I'm on the side of Bohr, no doubt.

TIPPETT: You've also written, "The old vision which Einstein maintained until the end of his life, of an objective world of space and time and matter independent of human thought and observation, is no longer ours. Einstein hoped to find a universe possessing what he called 'objective reality,' a universe of mountaintops which he could comprehend by means of a finite set of equations. Nature, it turns out, lives not on the mountaintops but in the valleys." Explain to me what you're describing there.

DYSON: If you look at the real nature, it's just so much more imaginative than a set of equations. What really happens in the universe is that nature finds all these extraordinarily complex structures which have their own rules. So, for example, the whole of biology is an example of that. Things happen in living creatures which you can't just describe with a set of equations. And that's true of most of science. That's true of chemistry and geology, of the whole of historical sciences.

TIPPETT: You say it's more like a rain forest than a mountaintop.

DYSON: Exactly. Exactly —that's exactly the metaphor. Complexity is the essence of things. So Einstein's universe of a sort of cold, hard space and time defined by a set of differential equations—it's there, but it's a very small part of the real universe. It's just the mountain peaks.

TIPPETT: But help me understand this. I think what's so intriguing — and we don't always think about it this way—is that the equations, the $E = mc^2$, that Einstein was laying out were not something that he was creating but discovering: equations, facts, rules, and principles, that somehow were there and undergird all of this. And that those equations and rules still somehow undergird this complex reality, the rain forest you're describing. Is that right?

DYSON: Yes. These equations are quite miraculous in a certain way. The fact that nature talks mathematics, I find it miraculous. I spent my early days calculating very, very precisely how electrons ought to behave. Well, then somebody went into the laboratory and the electron knew the answer. The electron somehow knew it had to resonate at that frequency which I calculated. That, to me, is something at a basic level we don't understand. Why is nature mathematical? But there's no doubt it's true. And, of course, that was the basis of Einstein's faith. Einstein talked that mathematical language and found out that nature obeyed his equations, too. Of course, his great moment was when they measured the deflection of light by the sun in 1919 and found that it followed his theory of gravitation.

TIPPETT: Was that the Eddington expedition?

DYSON: Yes, that was the expedition where Eddington made the observations and confirmed the theory.

TIPPETT: It did seem miraculous, didn't it, to people, that he was right?

DYSON: It was miraculous.

✳

In 1919, Einstein's theory of relativity was confirmed by two expeditions to Brazil and the West African coast to observe the total eclipse of the sun. The eminent British astrophysicist Arthur Eddington led the project. To the amazement of Eddington and the rest of the world, Einstein had correctly calculated that space could be distorted and light curved by gravity. Einstein was on the front page of newspapers worldwide. But when asked what he would have said had his theory not been proven correct by observation, Einstein replied, "I would have had to pity our dear Lord. The theory is correct all the same."

✳

DYSON: He had a marvelous sense of humor, and that's a very important part of life. The fact is that scientists have, on the whole, cultivated a sense of humor because so much of science is a history of failures. If you're a creative person, you know it's true in other kinds of creative life, but more so in science as so much of science ends up to be wrong. You do something, you spend weeks and months, and finally the whole thing collapses. You need to have a sense of humor, otherwise you couldn't survive. And Einstein, I think, understood that particularly well.

TIPPETT: I wanted to ask you what physicists are learning now that would befuddle Einstein, what would intrigue him. I suppose we've already wandered into that territory. What else is happening now that perhaps he made possible, but that might surprise him?

DYSON: Well, the big thing that he made possible, but which he never accepted, was black holes—places where big stars have collapsed and effectively disappeared from the universe, except that there's left behind a hole where the star used to be. So you have there a very strong gravitational field without any bottom. The black hole is the only place where space and time are really so mixed up that they behave in a totally different way. I mean, you fall into a black hole and your space is converted into time and your time is converted into space.

TIPPETT: Sort of the ultimate relativity?

DYSON: Yes. In a way, it's the most exciting, the most beautiful consequence of his theory. I mean, nature would not be the same without them. And I think if Einstein came back, he really would be surprised by that. If he came back now, he would have to accept that black holes are real and they're here to stay, and they are actually a tremendous triumph for his own ideas. It would be amusing to see his reaction. I'm sure he would accept it—and probably make some very suitable joke.

✳

Einstein's humor and humanity were revealed in his public appearances, but also in the vast correspondence he conducted with people of all walks of life. Here's a passage of a letter he wrote to one of his early biographers, who had asked Einstein to recall the details of receiving his first honorary degree. That happened, as it turns out, as part of the 350th anniversary of the founding of the University of Geneva by the Protestant reformer John Calvin.

> So I traveled there on the appointed day, and in the evening in the restaurant of the inn where we were staying, met some Zurich professors. I had with me only my straw hat and my everyday suit. My proposal that I stay away was categorically rejected, and the festivities turned out to be quite funny, so far as my participation was concerned. The celebration ended with the most opulent banquet that I have ever attended in all my life. So I said to a Geneva patrician who sat next to me, "Do you know what Calvin would have done if he were still here? He would have erected a large pyre and had us all burned because of sinful gluttony." The man uttered not another word. And with this ends my recollection of that memorable celebration.

If Albert Einstein can be said to have had a spiritual side, this expressed itself in part in his love of music. He played the violin from a young age and was a passionate concertgoer. He once mused that had he not been a physicist he would have been a musician. "I often think about music," he revealed. "I daydream about music. I see my life in the form of music." He carried his violin with him wherever he went.

*

Paul Davies is a theoretical physicist and cosmologist. I interviewed him from Sydney, Australia, where he spent fifteen years at the Australian Centre for Astrobiology, which he cofounded. He's currently at Arizona State University, where he is creating Beyond, a new center for fundamental concepts in science. Davies has written widely about Einstein's understanding of time and the intriguing scientific and existential questions it raises. Einstein referred to the human perception of time divided into past, present, and future as a "stubbornly persistent illusion." Before Einstein, science itself had taught society to think of time as a matter of fixed precision. Time was a universal constant, an arrow progressing at the same rate for everyone everywhere. Nineteenth-century notions of progress hinged on this belief about time. So did the modern Western concept of selfhood, of personal identity accumulated through the passage of time. But Einstein saw time as elastic, not absolute, curving and warping in response to space and mass and motion. I asked Paul Davies why this idea still sounds outlandish to a twenty-first-century mind.

PAUL DAVIES: The reason that people find Einstein's ideas weird is because we don't notice the effects that he discussed in daily life, and our brains have evolved their commonsense notions in order to cope with daily life. But we now have instruments of such extraordinary sensitivity that we can easily measure the warping of time just from everyday speeds. I suppose the one that is most dramatic is the global positioning system, without which, in Sydney at least, the taxi drivers would always get lost.

This system relies upon satellites which are orbiting the Earth, and if you don't factor in the warping effects of both motion and gravitation on time, you would very soon get lost, within minutes. And so this is an application of the theory of relativity.

TIPPETT: I think one of the most interesting stories you tell, as you describe what Einstein's contribution was to our understanding of space and time, is that, in fact, before Newton and Galileo, ancient cultures thought of time as organic and subjective and cyclical and part of nature. You say that the clock is an emblem of an intellectual straitjacket that was created in a relatively modern era by scientists, and that Einstein then restored time to its rightful place at the heart of nature. That's a very interesting idea.

DAVIES: It's certainly true that it was Galileo who recognized that time is the appropriate parameter in which to discuss the nature of motion and, in particular, falling bodies. And Newton then developed that idea into what is now sometimes called "the clockwork universe," that the entire cosmos is a gigantic clockwork mechanism slavishly following accurate mathematical laws to arbitrary precision. But it didn't enter into the practical world nearly so much until about probably the nineteenth century. The railroads were being established, and it was important for people to be at the station on time. And it was important to establish international time zones and national time zones for common ways of doing business. The telegraph was another very important step in establishing common time zones. It was curious that probably no more than a few decades after ordinary people began to be subjected to this temporal straitjacket, Einstein came along and upset the apple cart again. And I think historically part of the reason for this was that he was working in the patent office in Switzerland. Precision timekeeping and inventing clocks that would give ever-greater precision and enable time zones to be synchronized ever more accurately would have been something he would deal with on a daily basis.

TIPPETT: Right. He was in the capital of clocks, I guess, in Switzerland.

DAVIES: That's right. And so he was obviously thinking very much about measuring time, and this is what led him to the notion that your time and my time might appear different. We might measure different time intervals between the same two events if we're moving differently. And also your gravitational circumstances. Gravity slows time. Time runs a little bit faster on the roof than it does in the basement. We don't notice it in daily life. When you go upstairs and come down again, you don't notice a mismatch, but you can measure it with accurate clocks.

TIPPETT: From a religious perspective, there's something intriguing, though, in how these ideas of physics might seem to echo spiritual notions that you can find in Eastern and Western religious thought. In Australia—you're speaking from Australia—there's the notion of Dreamtime. There do seem to be echoes of that, of a sense of time as larger and malleable and mutable and not captive to human reality.

DAVIES: It's true that the Australian aboriginal concept of "the dreaming" reflects the perception of time of many ancient cultures, the notion that in a way there are two times. There's the one that we live our lives by on a minute-by-minute basis. But then there's this abstract notion, which is—maybe time is the wrong word. Maybe it's the opposite of time. Maybe it's eternity. This is a dualism, I think, that runs through all cultures, that there is time and then there is eternity, and that some things, in some sense, have an existence outside of time. They are eternal. And I don't fully understand, can't really grasp the aboriginal concept of the Dreamtime, but I think it will come closer to the Christian notion of eternity than it does to day-to-day, temporal sequence. I've been inspired by the work of Augustine, who lived in the fifth century and wrote extensively about the nature of time. And where I think he was spot on and where it resonates with Einstein has to do with the origin of time, the fact that time may have come into existence with the beginning of the universe. We think now that the universe began with a big bang, and people are fond of asking what happened before the Big Bang.

TIPPETT: That was also a legacy of Einstein, that we could discern that, correct?

DAVIES: Einstein gave us the so-called general theory of relativity in 1915, on which the notion of the expanding universe is based, and by extension the universe beginning with a so-called Big Bang. We know this is now 13.7 billion years ago. Einstein's theory of relativity says this was the origin of time. I mean, there's no time before it. And Augustine was onto this already in the fifth century because he was addressing the question that all small children like to ask, which is, "What was God doing before he created the universe?" Augustine said that the world was created with time and not in time. He placed God outside of time altogether, a timeless, eternal being. So we're back to eternity.

In 1930, Albert Einstein published an essay on religion and science in the *New York Times Magazine*. It was quoted and reprinted around the world. Einstein described his understanding that emotions such as longing and pain and fear gave rise to primitive forms of religion. Later he wrote that moral impulses drove what he called "the religions of civilized peoples, especially of the Orient." Einstein described his own inclination towards another kind of religious sensibility, which he called a cosmic religious sense. He discerned kindred glimpses of this feeling in such diverse figures as the prophets and psalmists of the Hebrew Bible, St. Francis of Assisi, and the Buddha. Writing in the *New York Times*, he noted:

> It is very difficult to elucidate this feeling to anyone who does not experience it. The individual feels the vanity of human desires and aims and the nobility and marvelous order which are revealed in nature and in the world of thought. Individual existence strikes him as a sort of prison, and he wants to experience the universe as a single, significant whole. The religious geniuses of all ages have been distinguished by this kind of religious feeling. In my view, it is the most important function of art and science to awaken this feeling and keep it alive in those who are receptive to it.

Paul Davies has written that theology was the midwife of science. In 1995, he won the Templeton Prize for Progress in Science and Religion, but like Albert Einstein, he's not a traditionally religious person. At the same time, like Einstein, he speaks often of God and especially of the mind of God. So I asked Davies what a physicist understands in using that phrase. And did Einstein's discoveries influence a new understanding for our time?

DAVIES: You have to understand how science emerged in Western culture, under twin influences—first of Greek philosophy, which taught that human beings can come to understand their world through rational reasoning. And then the second tradition began with Judaism, the notion of a creative world order, that there is a supreme lawgiver who brought the universe into existence at a finite time in the past and orders the universe according to a rational plan. Both Christianity and Islam adopted this linear time and a creative world order, and the scientists had that tradition. They said, "Well, there's an order in nature, but it's hidden from us." We don't see it in daily life. We have to use arcane procedures of experiment and mathematics to uncover this what I like to call mathematical code that underpins nature. We now call that the laws of physics. But there emerged this notion that human beings could come to understand it, could read the mind of God, because the application of human reasoning and human inquiry was a tremendous thing. And the birth of science can be identified with this step.

TIPPETT: I do hear echoes of Einstein also in that kind of language. Here's something he said in 1955: "I want to know how God created this world. I'm not interested in this or that phenomenon, in the spectrum of this or that element. I want to know His thoughts. The rest are details."

DAVIES: Einstein was fond of using the word "God," and there are many famous quotations. "God does not play dice with the universe" is his antipathy to quantum physics and its indeterminism. Sometimes he was really using God as just a sort of *façon de parler*, a convenient metaphor. But he did have, I think, a genuine theological position. He did not believe in a personal God. He made that very clear. But he did believe in a rational world order, and he expressed what he sometimes called a "cosmic religious feeling," a sense of awe, a sense of admiration at the intellectual ingenuity of the universe. Not just its majesty, its grandness, its vast size, but its extraordinary subtlety and beauty and mathematical elegance—something that people who are not physicists find very hard to grasp. But to a professional physicist, this notion of an underlying mathematical beauty is part and parcel of the subject.

TIPPETT: And you also raise religious, theological questions that, for you, still flow out of these great discoveries of Einstein and of physics as we know them now, the burning questions that remain. Maybe we don't need God for the laws of physics to do their job, but where do the laws of physics come from? Why these laws rather than others? And here's some language of yours. "Why a set of laws that drive the searing, featureless gases coughed out of the Big Bang toward life and consciousness and intelligence and cultural activities such as religion, art, mathematics and science?" Are those questions that you can keep asking now, this far down the road? Did Einstein consider questions like that?

DAVIES: For me the crucial thing is that the universe is not only beautiful and harmonious and ingeniously put together, it is also fit for life. And physicists have traditionally ignored life. It's too hard to think about. More and more, though, I think we have to recognize that if the laws of physics hadn't been pretty close to what they are, there would be no life. There would be no observers.

Now, sometimes we just shrug and say, "Well, so what." You know, "If it had been different, we wouldn't be here to worry about it." But I think that's unsatisfactory. And the reason it's unsatisfactory is because the universe has not only given rise to life, it's not only given rise to mind, it's given rise to thinking beings who can comprehend the universe. Through science and mathematics, we can, so to speak, glimpse the mind of God, as we've been discussing.

And I think that this suggests, to me anyway, that life and

mind are not just trivial extras. They're not just an embellishment on the grand scheme of things; they're a fundamental part of the nature of the universe. And if you imagine playing the role of God and having some sort of machine in front of you with a whole lot of knobs, and you twiddle the knobs and change things—twiddle one knob, make the electron a bit heavier; twiddle another knob and make the strong nuclear force a bit stronger—you soon discover that you have to fine-tune those settings to extraordinary precision in order for there to be life. And the question is, what are we to make of that? And really these things, at the end of the day, boil down largely to a matter of personal choice, because we can't really test either. Or certainly not in our current state of knowledge.

Paul Davies points out that the current conversation between science and religion is different in physics than in biology. So when he speaks of the fine-tuning of the universe, or when Einstein spoke of a "mind" or "superior spirit" behind nature, this does not mirror the biologists' debate between Darwinian evolution and intelligent design. The order behind the universe that Einstein discerned was manifest in the laws of physics. Still, though Einstein rejected the notion of a creator who would interfere with the laws ordering his own creation, his discoveries did give rise to the fields of quantum physics and chaos theory. And some scientists in those fields are now suggesting that there might be room for an involved God within the laws of physics themselves. I asked Paul Davies about this.

DAVIES: Yes, there has always been this problem for physicists about an active God. If God does anything, God has to be at work in the world. Now if we go back to the sort of universe that Newton had and the one that Einstein supported, the notion of a deterministic universe, a clockwork universe, this becomes a real problem. Because if God is to change anything, then God has to overrule God's own laws. And that doesn't look like a very edifying prospect theologically or scientifically. It's horrible on both accounts.

But when one gets to an indeterministic universe, if you allow quantum physics, then there is some sort of lassitude in the operation of these laws. There are interstices having to do with quantum uncertainty into which, if you want, you could insert the hand of God. So, for example, think of a typical quantum process as being like the roll of a die. "God does not play dice," Einstein said—well, it seems that God does play dice. Then the question is, if God could load the quantum dice, this is one way of influencing what happens in the world, working through these quantum uncertainties. Some people certainly have pushed that idea. John Polkinghorne is one who's spoken about it. Bob Russell of the Center for Theology and Natural Sciences in Berkeley likes that point of view of God not in any sense usurping the laws of physics, but working within the inherent lassitude that quantum physics provides. It's a possible way of God to gain cause or purchase in the world without changing any of the laws that we know.

TIPPETT: I think, as we close, I'd like to come back to this idea of eternity. We touched on that a bit when we were talking about time—which was such an important subject for Einstein—and this idea that is in many cultures and many religious traditions of a distinction between the temporal and the eternal. I'd like to read you a passage from a letter that I found that Einstein wrote when he was a bit older and just see how you respond to it as a physicist. He wrote this actually to the queen of Belgium, who was grieving a great loss. And he said to her, "And yet, as always, the springtime sun brings forth new life, and we may rejoice because of this new life and contribute to its unfolding. And Mozart remains as beautiful and tender as he always was and always will be. There is, after all, something eternal that lies beyond the hand of fate and of all human delusions. And such eternals lie closer to an older person than to a younger one, oscillating between fear and hope. For us there remains the privilege of experiencing beauty and truth in their purest forms." I don't think this is an Einstein many of us know when we just think of his scientific legacy.

DAVIES: Those are beautiful words, and I was quite unaware of them, very poetic. And I can see where they're coming from because, as we discussed earlier, Einstein was the person to establish this notion of what is sometimes called block time—that the past, present, and future are just personal decompositions of time, and that the universe of past, present, and future in some sense has an eternal existence. And so even though individuals may come and go, their lives, which are in the past for their descendants, nevertheless still have some existence within this block time. Nothing takes that away. You may have your threescore years and ten measured by a date after your death. You are no more. And yet within this grander sweep of the timescape, nothing is changed. Your life is still there in its entirety.

TIPPETT: It's a wonderful thought, isn't it? It opens up our imagination about what it means to be human and the universe, our place in it.

DAVIES: I think that physics impacts upon our view of the universe and our place within it in so many ways—in the nature of time, in the nature of reality through quantum physics, and, as we've discussed, in the fact that the universe is fit for life and that we're a component in this biofriendly universe that has such ingenious laws that can enable life to come into existence. And this puts our own position on this planet into a very different context. It cuts both ways, because on the one hand we can see that we're not the center of the universe, we're not the pinnacle of creation, that we are maybe a small part, maybe only one among myriad living systems throughout the universe. And yet, nevertheless, we have emerged. And we can truly feel part of nature in a cosmic sense, not just in a local sense, but in a genuinely cosmic sense. I think that's deeply inspiring whatever one's religious convictions, and even if you have no religious convictions. I often say that if I talk to someone like Steven Weinberg, who's a professed atheist and quite militantly so . . .

TIPPETT: He's the one who said, "The more we learn, the more pointless it seems"?

DAVIES: That's right, and yet, nevertheless, he will share in the awe, the wonder, the majesty, the beauty of the universe in this cosmic connection that I've been talking about. He sees the same facts as I do but can't bring himself to believe that there's any point behind it all. And that's where he and I will part company. We'd agree on all of the science, but to me it overwhelmingly suggests that the universe is about something, that there is a point to it, and that we're part of whatever point that is.

*

Here in closing are some lines from a letter Albert Einstein wrote in 1927:

I cannot conceive of a personal God who would directly influence the actions of individuals or would sit in judgment on creatures of His own creation. I cannot do this in spite of the fact that mechanistic causality has, to a certain extent, been placed in doubt by modern science. My religiosity consists in a humble admiration of the infinitely superior spirit that reveals itself in the little that we, with our weak and transitory understanding, can comprehend of reality. Morality is of the highest importance, but for us, not for God.

2

✳

The Spirit as an Emergent Life Force

"THE BIOLOGY OF THE SPIRIT"

*I*n the summer of 2005, a few colleagues and I went to the Chautauqua Institution's week on "the brain." We were invited by Chautauqua's Religion Department, which was focusing that week on the nature of love. Those two subject areas might seem on the surface to deal with distinctly separate realms of human reality—reason and emotion. But one of the lessons of the week was that modern science is turning up an intricate and fascinating interrelationship between them. What we are learning about "the three-pound human brain," as Sherwin Nuland likes to refer to it, may compel us to reconcile Western civilization's split between body and spirit.

Sherwin Nuland first translated his personal knowledge of human physiology into literature with his award-winning bestseller, *How We Die*. He epitomizes a phenomenon I've observed among scientists of many disciplines: he is possessed of a passion for his subject borne, perhaps counterintuitively, of rational observation and scientific expertise. "I want everyone to know what I have come to know," he wrote in the introduction to his 1997 book *The Wisdom of the Body*.

After he had chronicled processes of human death in *How We Die*, he delved into the primary processes that support and sustain human life—how we live. Years of treating disease in the body left him in awe, above all, of the fact of health as a norm. Knowing what can go wrong, he says, has given him a tremendous respect for much more that goes right, moment to moment.

He chose as the epigraph to *The Wisdom of the Body* this thought of St. Augustine:

> *Men go forth to wonder at the heights of*
> *mountains, the huge waves of the sea,*
> *the broad flow of the rivers, the vast*
> *compass of the ocean, the courses of the*
> *stars: and they pass by themselves*
> *without wondering.*

But "wonder" for St. Augustine was a religious experience that drove back to a creator. Dr. Nuland looks within the body not only for the source of his wonder but for the driving force of his capacity for wonder itself. He makes the provocative suggestion that what we call the human spirit—our capacity for beauty and love, our drive to create balance in life and moral order in society—is an evolutionary accomplishment of the most complex organism on the planet, the human brain. Within our very beings, he says, we sense the threat of chaos, and we sometimes yield to it. But overwhelmingly, individually and collectively, we seek balance. We transcend mere impulse and reason. Sherwin Nuland has given himself over to charting transcendence rooted in flesh and blood and bone, DNA and neurotransmitter and enzyme.

I couldn't help noticing as I traced the line with Sherwin Nuland that I trace with all of my guests—the intersection of large ideas with concrete experience—how his personal story mirrors the development of his thinking.

He had a difficult early life, which he describes in his autobiographical work, *Lost in America*. In his forties, he almost succumbed to a grave spiral of clinical depression. He lost the Orthodox Jewish faith of his childhood, but gained an animating faith in human realities that he had spent his career exploring. One of his favorite quotes, attributed to Philo of Alexandria, has now become one of my favorite quotes: "Be kind, for everyone you meet is fighting a great battle."

Sherwin Nuland's ideas might richly inform many religious perspectives; and as he admits, they do not rule out the idea of a creator. We have an interesting exchange about how his concept of the spirit as emergent in human life and relationship corresponds intriguingly, in fact, with the Hebrew biblical word for "soul"—*nophesh*, in English transliteration. His idea of our bodies evolving our spirits could also be heard as parallel to the suggestion of the physicist/theologian John Polkinghorne, who believes in a God who created something more beautiful than a ready-made world—a world with an inborn capacity to become and create itself.

∗

The Biology of the Spirit

KRISTA TIPPETT, host
SHERWIN NULAND, surgeon and author

∗

In *How We Die* and all of his later works, Sherwin Nuland reflects on the meaning of life by way of scrupulous and elegant detail about human physiology. A clinical professor of surgery at Yale University, he also teaches bioethics and medical history. He was a practicing surgeon for thirty years, treating upwards of ten thousand patients. After the success of his book about death, Dr. Nuland turned his attention to the infinite variety of processes that maintain human life moment to moment. He finds the source of his greatest wonder within the body, and he's increasingly fascinated these days by what new understanding of the brain can reveal about what makes us human. He believes that human beings, alone among living creatures, became aware over time of the cost of decay in the world around them and of their own finitude. "In response," he writes, "our brains developed a capacity for spirit, for seeking lives of integrity and equanimity and moral order."

Dr. Nuland's own story mirrors this progression. He relinquished the Orthodox Jewish faith of his immigrant upbringing as he recovered from a grave struggle with clinical depression in his forties. But precisely through that experience and through years of treating disease in the body, he became fascinated with the biological roots of human health and spirit.

NULAND: When one has spent that amount of time studying abnormalities, one develops an enormously healthy respect for normal. One develops an enormously healthy respect for the equilibrium that is maintained in order for us to continue in healthy life—especially when you're in surgery, and you look inside an abdomen and realize how many things could go haywire and yet they don't. Or you look at the structures that you're dealing with and recognize that everything is just humming along beautifully, nobody is running it. It's just going by itself and the orders are coming from somewhere. This little bit of pathology that you're taking care of is in the corner somewhere. Now, we know it affects the whole body. But when you look at it, you're most impressed with the normal structures.

TIPPETT: You also had a serious experience of depression, which you wrote about in a more recent book, *Lost in America*. I'd also be curious about how that experience formed your interest, your fascination with your physical self, with the body.

NULAND: When I became depressed, I came very quickly to admit something to myself that I had really been aware of on some level but refused to come out and say directly—which was that my religious beliefs, what I thought were my religious beliefs, were nothing more than obsessional thinking. It really had to do with fear of—whether you want to call it hellfire or punishment of some kind.

TIPPETT: And these would be the beliefs of this Orthodox Jewish upbringing?

NULAND: How does one know? Certainly I, since that time, have met many, many, many Jews, just as I've met many Catholics and Protestants who have deep faith, who really believe, and obsessional thinking has nothing to do with it, and fear of punishment has nothing to do with it. This is true faith. My problem was me and not what Judaism represents. It was clear to me that the behaviors that I was exhibiting, that I thought were in keeping with Jewish religious precepts, had to do with superstitious fears of punishment. And if I was going to get out of this depression, I was going to have to give that all up. And I just did it. I did it by an act of will.

TIPPETT: Since that time, is there a connection between what you gave up in terms of these obsessional thoughts, the religious ideas that weren't good for you, and what you began to think about in their place? Because you didn't really stop thinking about what it means to be human or the fact that there's a human spirit, which are often connected with religious beliefs.

NULAND: That's right. It's the human spirit that got me through. It was the sense that there is a richness in this world that's enormous fun if you can find it, and it's the kind of fun that you can have while actually making the world a better place for other people, too. There's an integrity to it in the sense of a oneness, of a unity. If you are discovering the essence of what it means to be human, you are freeing yourself from all these enmeshings and thinking about yourself. You're really thinking about humanity and the human spirit and, accordingly, other people. And the sense of accomplishment when you accomplish something from that intellectual and emotional background is enormous because of this freedom.

I was shackled by neurotic thoughts, by essentially being so twisted in the meaning of what I was doing, the spiritual meaning from some supernatural being that I didn't really understand. Once that was gone, what a wave of freedom, what a liberating thing it was. It was as though I'd been bottled up and someone took the cork off the bottle. And luckily for me, it's continued to come out. When one gets tempted to take up the obsessions or neurotic symptoms again, one begins to think of not just the cost of doing that but what you would lose, the pleasure that you would lose, the rewards you would lose, the sense of self you would lose, the feeling of being a part of an open community that you would lose. I know this is a lot of abstraction.

TIPPETT: I don't think it's abstraction. I think a lot of people will know what you're talking about and will recognize themselves there. But again, what is fascinating is that you thought about the human spirit in a whole new way and your own spirit. And you did so also by connecting that with the work of your life, your work with the human body. Here is some language from your book: "The human spirit is the result of the adaptive biological mechanisms that protect our species, sustain us and serve to perpetuate the existence of humanity." That could sound like a kind of antiseptic and reductionist statement, but in fact . . .

NULAND: It could sound like a meaningless statement.

TIPPETT: But you make it and you surround it with a sense of great wonder.

NULAND: Well, you just got the word. I've been sitting here on the edge of my seat, hoping, When am I going to get to say this word, "wonder"? Wonder is something I share with people of deep faith. They wonder at the universe that God has created, and I wonder at the universe that nature has created. This is a sense of awe that motivates the faithful, motivates me. And when I say motivates, it provides an energy for seeking. Just as the faithful will say, "We are seeking," I am seeking.

We're seeking different things. I'm seeking an understanding of this integrity of everything, of this unity of everything, of the equilibrium of not just the *homeostasis*, as the physiologists say, the staying the sameness, but of the closeness that we are constantly coming to chaos. I have had chaos. I've had chaos to the point where I thought my mind was lost. This gives me a deeper appreciation of equanimity, not just continued existence but continued learning, continued productivity.

✳

Here's a passage from Sherwin Nuland's book *The Wisdom of the Body*:

> Notwithstanding the tragedies that humankind has visited on itself individually and collectively, and the havoc we have wreaked on our planet, we have become endowed nevertheless with a transcendent quality that expands generation upon generation, overcoming even our tendency toward self-destruction. That quality, which I call spirit, has permeated our civilization and created the moral and esthetic nutriment by which we are sustained. It is a nutriment, I believe, largely of our own making . . .
>
> As I define it, the human spirit is a quality of human life, the result of living, nature-driven forces of discovery and creativeness; the human spirit is a quality that *Homo sapiens* by trial and error gradually found within itself over the course of millennia and bequeathed to each succeed-

ing generation, fashioning it and refashioning it—strengthened ever anew—from the organic structure into which our species evolved so many thousands of years ago. It lives while we live; it dies while we die . . . It is neither soul nor shade—it is the essence of human life.

✳

NULAND: I think there is an evolutionary accomplishment of the human cortex, the cortex of the brain, and the way it relates to the lower centers of the brain and the way it relates to the entire body. The way it accepts and synthesizes information and uses information from the environment, from the deepest recesses of the body. The way it recognizes dangers to its continued integrity. And I think that this is precisely what the human spirit is doing. The human spirit is maintaining an equilibrium, largely related to its normal physical and chemical functioning.

TIPPETT: So there's a biological underpinning for intelligence that evolved over a great large span of time in human beings, but then we developed something else, which is consciousness. Is consciousness the same as spirit?

NULAND: I don't think so. Consciousness is only a kind of awareness of our surroundings, an awareness of our emotions, an awareness of our responses. The human spirit is something much greater. The human spirit is an enrichment. It's the way we use our consciousness to—I keep using this word—to synthesize something better than our mere consciousness, to make ourselves emotionally richer than we in fact are.

TIPPETT: Almost to transcend what was given.

NULAND: To transcend. And this is what I mean when I say that we are greater than the sum of our parts: that because of the trillions of cerebral connections we have, and the way our species for the past forty thousand years since modern Homo sapiens appeared on Earth has adapted to stimuli from the outside, we have relentlessly pursued this upward course, I believe, toward creating the richness of the human spirit.

There is a word that the neuroscientists use when talking about why a certain series of circuits or group of circuits in the brain is activated. That word is "value." There are pathways in the brain that have survival value. So when a stimulus comes in and the brain has fifty thousand different ways of responding to it, some of those are useful for survival. Some of those will either prevent survival or mar survival. And the human brain, in classical evolutionary pattern, will pick the one that is healthiest, that gives greatest pleasure. What gives greater pleasure than a spiritual sense? So I think of this as natural selection in an emotional form, and I think it is almost like choice. Because when you're talking about selection in the brain, there are processes of choice. The brain has a way of evaluating what is best for the organism. And what is best for the organism is not just survival and reproduction but beauty, an aesthetic sense.

TIPPETT: Okay. So we human beings have chosen to value beauty.

NULAND: You bet. We've chosen. Now, it's an unconscious process, but what we know about unconscious processes are that for every conscious process there are eight million zillion trillion unconscious ones, and they are in fact what will eventually determine what's conscious and what we can understand. So again, to reiterate, this is a process of natural selection. A stimulus comes in. There are many, many ways of responding to it. Some of those ways are counterproductive, some are kind of ordinary, and some really give satisfaction and enhance the richness of our lives. And without knowing it, our circuits are choosing those, and this is what I mean by the human spirit.

TIPPETT: And I'm so struck by the loving detail with which you describe these trillions of circuits and what is happening, how amazing those biological mechanisms are, and how amazing you find it to be.

NULAND: It's wonderfully exciting. Here are these seventy-five trillion cells, and every cell has hundreds of thousands of protein molecules in it. They are constantly interacting with one another in what would appear to be chaos. In fact, if you were to be able to lower yourself into a cell, you'd be terrified because it would seem so chaotic. If it had sound, you couldn't live with it, it would be so noisy. And yet what is actually occurring is that these reactions are all counteracting threats to the survival of that cell. And I think that there is within the human organism—and only the human organism because of our cortex and our ability to process information—there is an awareness of the closeness of chaos.

There's a lot of evidence for that, including cultural evidence. I talk often about the polarity of our thinking. We talk, for example, about good and evil. One of my favorite examples of this I got from my Orthodox Jewish background, which is the principle of the good inclination living in balance with the evil inclination, that one must make that choice at all times. Now, the Greeks, who expressed it as chaos versus cosmos, they had a sense that there was an order up there in the universe, but we live in chaos. And listen to the popular music that people in their teens and twenties and thirties are listening to, and if you listen carefully you're always going to hear the heartbeat in the background.

TIPPETT: You're saying that we are always seeking rhythm.

NULAND: Harmony, order, integrity in the sense of oneness. And this is why monotheism is so pervasive. Everybody says, "Oh, this is wonderful, the Jews discovered monotheism, the Christians embrace it, the Muslims embrace it," as if this has to be the right thing. Why does it have to be the right thing? Why is this better than a bunch of polyglot gods or polymorphic gods? It's because we need unity, predictability. We need a moral sense. We need a moral sense to prevent the chaos that somehow we recognize we are living close to.

TIPPETT: I'm also thinking of the beginning of Genesis, of the beginning of the Hebrew Bible, where the original creative act is creating order.

NULAND: That's right. They were responding to precisely that same deep awareness that hadn't even reached the level of the unconscious mind yet that I'm talking about.

TIPPETT: The spirit, for all of its wonder and the good that we associate with it, also has base qualities and has a dark side. How do you think about that in this scheme?

NULAND: I think it has to do with the nearness of chaos, which is always a temptation. It's like the butterfly and the flame. We are tempting ourselves with evil, we are tempting ourselves with that which is destructive, and we to some extent succumb to it.

If you talk to psychoanalysts about severe neurotic disease, they often talk about the personality that skates to the edge and then rescues itself from the edge. We are so tempted to go to hell with ourselves, as it were—that's a theological expression—that we actually do come near it, even recognizing the other pole. And this is what the Greeks meant when they were talking about Eros versus Thanatos, the love and life sense against the death sense. I don't think it's in very many of us to deliberately choose destruction, but we play with it and it licks us and burns us and can ruin lives. So this is all part of that polarity that I was talking about, the fear of chaos, which makes us look for order.

✳

Here's another passage from Sherwin Nuland's writing:

> Always the purpose of treatment is only to restore nature's balance against disease. There is no recovery unless it comes from the force and fiber of one's own tissues. The physician's role is to be the cornerman—stitch up the lacerations, apply the soothing balm, encourage the use of the fighter's specific abilities, say all the right things—to encourage the flagging strength of the real combatant, the pummeled body. As doctors, we do our best when we remove the obstacles to healing and encourage organs and cells to use their own nature-given power to overcome.
>
> We have always known this. Every system of so-called primitive medicine I have ever encountered views disease as the imbalance of certain factors whose proper interrelationships must be reestablished if recovery is to take place . . .
>
> I have spent the adult years of my life being nature's cornerman. I have provided it with whatever boost was

needed, cheered it on, and felt the exhilaration of watching its formidable powers wheel into action once I have helped remove the impediments. An inflamed organ is excised, an obstruction is bypassed, excessive hormone levels are reduced, a cancerous region is swept clean of tumor-bearing tissue—and the wrongs are redressed, thus allowing cells and tissues to take over the process of reconstituting equilibrium. Surgeons are no more than agents of the process by which an offending force may be sufficiently held at bay to aid nature in its inherent tendency to restore health. For me, surgery has been the distilled essence of W. H. Auden's perceptive précis of all medicine: "Healing," said the poet, "is not a science but the intuitive art of wooing nature."

❋

TIPPETT: I don't know if you use the word "soul." Would you use the word "soul" and "spirit" interchangeably?

NULAND: I don't think I ever use the word "soul." "Soul" has implications that I'm trying to stay away from.

TIPPETT: I've been thinking, there is in Jewish tradition the *nephesh*, the soul that is emergent, that is quite different from, say, the Christian idea of the soul. There is a Jewish sensibility of the soul as being something that emerges in relationship. And I do think that there is some affinity between that image and the way you imagine the human spirit to be this evolving work of humankind.

NULAND: Well, you have just told me something I've got to think a lot about, because it had never occurred to me. But what

I've got to do is think of the theological implications of *nephesh*, because I suspect that I know far more than I think I know, just as we all know far more than we think we know. We know all these words and if we were to sit down with ourselves in a quiet room or just sit with a pencil, extraordinary things would come out. And they would be correct. So I've got to sit in the corner, and I've also got to talk to some better theologians than I about the implications of *nephesh*, because I assume—why does this hit me so hard? Because I assume that I know things about that concept that I don't realize I know.

TIPPETT: Well, it may have been something you breathed in in that Orthodox Jewish air that you grew up in.

NULAND: It all is related, you know, to the Greek notion of *pneuma*, the notion that the soul exists in the universe and with your first breath you inhale the *pneuma*, P-N-E-U-M-A, and that is the life-giving force. "Pneuma" is actually etymologically related to "psyche." So you get psyche, spirit, soul all together in one, but the origin of it is this thing that you inhale. So all of these traditions end up going back, I think, to something that all Homo sapiens share. And if all Homo sapiens share it, one of two things has to be true: either God gave it to everybody, or it's a universal on some level of awareness, it's in our DNA. I choose to think it's biological.

✳

Sherwin Nuland reflects on meaning by way of biological struggles he's witnessed in medicine and in life. In his book *How We Die*, he explored his knowledge of physical pathologies and their infinite variety in the context of human stories. He included personal mem-

ories of the aging and death of his grandmother, in Yiddish, his "Bubbeh." He shared a room with her for eight years of his childhood, in
a Bronx apartment that housed three generations. He wrote:

> It must have been after my mother died that I first began
> to be conscious of just how ancient Bubbeh was. Since ear-
> liest memory, I had amused myself from time to time by
> playing idly with the loose, unresilient skin on the backs
> of her hands or near her elbows, gently drawing it out like
> stretched taffy, then watching in never-lessening wonder
> as it slowly resettled into place with an easy languor that
> made me think of molasses . . .
>
> Bubbeh moved from room to room in slippered feet
> and with great care. As the years passed, the walk became
> a shuffle, and finally a kind of slow sliding, the foot never
> leaving the floor . . .
>
> Slowly, her vision, too, began to fail. At first, it be-
> came my job to thread her sewing needles, but when she
> found herself unable to guide her fingers, she stopped
> mending altogether, and the holes in my socks and shirts
> had to await the few free evening moments of my chron-
> ically fatigued aunt Rose, who laughed at my puny at-
> tempts to teach myself to sew. (In retrospect, it seems
> hardly possible that one day I would be a surgeon; Bubbeh
> would have been very proud, and very surprised.) After
> some years, Bubbeh could no longer see well enough to
> wash dishes or even sweep the floor because she couldn't
> tell where the dust and dirt were. Nevertheless, she
> wouldn't give up trying . . .
>
> In my early teens, I saw the last traces of the old com-
> bativeness disappear and my grandmother became almost
> meek. She had always been gentle with us kids, but meek-

ness was something new—perhaps it was not so much meekness as a form of withdrawal, an acquiescence to the expanding power of the physical disablements that were subtly increasing her separation from us and from life.

✳

TIPPETT: You tell so many wonderful stories in all your writing. You chose to write about your grandmother, your Bubbeh, in *How We Die*. You wrote about her death, and so many of the stories you tell are about moments in surgery or in hospitals and individual lives in the balance, and they're all so unique. You wrote about your grandmother and you've talked about how many letters you got about that from people all over the country. You quoted this pig farmer in Iowa who wrote, "Thank you so much for sharing your beloved Bubbeh with us. I now love her too, as I have known her by another name in another time in another place."

NULAND: Exactly.

TIPPETT: It struck me, this paradox of how different we all are in every one of our situations, in living and dying, and yet how alike.

NULAND: Do you know what I learned from writing that book, if I learned nothing else? The more personal you are willing to be and the more intimate you are willing to be about the details of your own life, the more universal you are.

TIPPETT: Isn't that interesting?

NULAND: Isn't that interesting. And when I say universal, I don't mean universal only within our culture. You know, there's a lot of balderdash thrown around: "You don't understand people who live in Sri Lanka and their response to the tsunami because you just don't know that culture." Well, there's an element of that. But to me, cultural differences are a kind of patina over the deepest psychosexual feelings that we have, that all human beings share, that they share by virtue of the physical properties of their body and the kind of brain that they have, which bring out certain sorts of strivings, certain sorts of emotional needs that are indeed universal.

TIPPETT: And how do we make use of that knowledge? Or do we just know it?

NULAND: I think we make use of that knowledge to perpetuate love. There is a book that I wrote called *Lost in America*, and there is a quotation in that book. It's the epigraph of that book, and it's attributed to Philo. Nobody who's a Philo expert has been able to find it for me, and I certainly can't find it: "Be kind, for everyone you meet is carrying a great burden." Well, that's the philosopher's stone.

When you recognize that pain and response to pain is a universal thing, it helps explain so many things about others, just as it explains so much about yourself. It teaches you forbearance. It teaches you a moderation in your responses to other people's behavior. It teaches you a sort of understanding. It essentially tells you what everybody needs. You know what everybody needs? You want to put it in a single word? Everybody needs to be understood. And out of that comes every form of love. If someone truly feels that you understand them, an awful lot of neurotic behavior just disappears. Disappears on your part, disappears on their part. So if you're talking about what motivates this world to continue existing as a community, you've got to talk about love. And you can't talk about—oh, I'm going to get into hot water for this—you can't talk about this phony concept of love that so many of the religious throw around based on God's love. You've got to think about this in terms of human biology, including emotional biology.

TIPPETT: Love is such a watered-down, wishy-washy word in our culture.

NULAND: Well, sure. It's misused, it's bastardized. And it becomes somebody's slogan.

TIPPETT: But if you approach everyone, as you say—I love this epigraph, "Be kind, for everyone you meet is fighting a great battle . . ."

NULAND: That's it, "fighting a great battle." Yes, that's even better.

TIPPETT: That also engenders the qualities you spoke about: patience, hospitality, compassion—virtues that are at the heart of all the great religious traditions, right?

NULAND: Of course. There's the universal again. And my argument is it comes out of your biology because on some level we understand all of this. We put it into religious forms. It's almost like an excuse to deny our biology. We put it into pithy, sententious aphorisms, but it's really coming out of our deepest physiological nature.

TIPPETT: You're very clear that some people could read the way you describe reality and even your sense of spirit as something that has evolved—and could also be religious and hold that together with a sense that there's a God.

NULAND: Of course. These are two different belief systems. There isn't a reason in the world that the religious have to explain their faith on a scientific basis. What is needed between science and religions is not a debate but a conversation, each one saying, you know, "You're here to stay and I'm here to stay, so let's find out how our relationship can be of greatest benefit to this world." And in my book on Maimonides, I pointed out that this debate between the two did not exist until the philosophers of the enlightenment created that debate.

TIPPETT: Maimonides was a physician and a philosopher . . .

NULAND: Yes. And a theologian.

TIPPETT: . . . and a theologian.

NULAND: Aquinas was a philosopher and a theologian. Averroës was a physician, a philosopher, and a theologian. All three of these people knew essentially all the knowledge available to anybody at that time. And they were engaged in the pursuit of bringing science and philosophy—and specifically Greek philosophy—on the one hand together with faith on the other hand. That's what they did. When people talk about Galileo, they say, "Oh, my God, Galileo, he was a heretic." Not at all.

TIPPETT: Not true, I know. It's just bad history.

NULAND: Galileo's entire pursuit was to bring his theory into keeping with church doctrine.

TIPPETT: When I was reading your writing about the evolution of spirit, of humanity as a creation of humankind and a creation

of the brain, I was thinking of John Polkinghorne, who's a British physicist and theologian. He believes in God, but he says God did something much more clever than create a clockwork world. God created a world that could make itself.

NULAND: There it is, you know.

TIPPETT: But I'm not saying there's any reason to force that either.

NULAND: And you know what else God did? Let's say categorically we're both people of faith. He gave humankind free will, and free will becomes the essence of the whole thing. And not just free will in the conscious sense, but he would have created the free will that makes the synapses and the nerve cells and the neurotransmitters, allows them to make choices. And given the opportunity to make choices, they will always choose the more—let's use that big word—*salubrious* way. Salubrious in the classical sense of healthy, physically healthy, emotionally healthy, the thing that's going to make it survive most likely and provide it with the most pleasure. And the moral sense provides people with more pleasure than anything. That's been my experience, that a sense of oneself as a good person whose life isn't sacrificed for others but is based around community and love gives one a sense of self that is the greatest pleasure that anybody can have. We say virtue is its own reward. It's a little homily, but there's a lot behind that homily. Every cliché has a reason.

TIPPETT: This adventure of learning about the brain, which you are steeped in—I wonder how you think this kind of knowledge will begin to reach ordinary people. Are there ways it will turn up in our culture at a more basic level?

NULAND: What happens to science—or what has happened to science since the great discoveries began to be made around the middle of the twentieth century—is that increasingly the public, by reading about some of these technological events and phenomena over and over again, it starts seeping in. Bit by bit people began to understand DNA. In 1955, just two years after Watson and Crick did their nice thing and published that great paper, nobody could figure out what DNA was. Now everybody knows what DNA is.

TIPPETT: Right.

NULAND: Also all kinds of notions of heredity and genetics. When people first started talking about stem cells and cloning, they were a mystery except in some sort of comedic sense. But bit by bit, people are recognizing scientific observations and discoveries. And my guess is that neuroscience, as it evolves, will slowly become something apprehensible to most reasonably well-educated people. Of course, it would help if we had a better way of scientifically educating ten-year-olds, but we are not in that situation in this country.

TIPPETT: But you know, the neuroscience that you're describing is also something that I think people could use to make their lives more fulfilling. That kind of knowledge is also a form of power.

NULAND: I like to think that if people really understood the way their brains work, they would be as overwhelmed with wonder as some of us are. I like to think that they would have a completely different sense of the human organism and its potentialities and would try to live up to its greatest potentialities.

3

*

Discovering the Globalization of Medicine

"HEART AND SOUL"

Mehmet Oz is one of the most respected and dynamic of a new generation of doctors taking medicine to new spiritual as well as technological frontiers. He is well known these days as a television and publishing personality. But he continues as director of the Cardiovascular Institute at Columbia University. When I interviewed him in 2004, he had published his first semi-autobiographical book, *Habits of the Heart*. And he was renowned in medical circles, if not beyond them, for patenting several medical technologies that had dramatically improved the prospects for cardiac patients facing heart transplantation or death. At the same time, he had introduced meditation, prayer, reflexology, acupuncture, yoga, and massage into the operating theater and recovery room.

My conversation with Mehmet Oz made me wonder why up to now we have not had a rich public vocabulary for discussing an integrative approach to medicine. This remains true even as "alternative" medical approaches are offered widely in major hospitals and medical schools across the country. "Alternative" sounds like something off center, not broadly accepted or esteemed. The other widely used phrase, "traditional medicine," summons up visions of comforting old practices that technological sophisticates have surely outgrown.

Mehmet Oz prefers to speak of "global medicine" and in doing so he puts these trends in spacious perspective. He suggests that integrative medicine is the realignment that global networks, communications, and travel have brought to medicine, just as they are realigning every other human endeavor from business to popular culture. He sees integrative medicine as a mutually enriching encounter of best practices from Western and Eastern cultures. In thinking this way, Oz does not belittle the radical advancements that Western medicine has made. He confesses his gratitude to earlier generations that innovated the lifesaving surgical techniques he performs routinely. But he finds, to his astonishment—and, one senses, delight—that other traditions and philosophies of healing often work precisely at the boundaries of our most advanced techniques. And he is not working merely to rid his patients of disease, but to help them be well.

I'm intrigued by the expansive definition of spirituality that I encounter when I speak with Mehmet Oz and other younger physicians. They refer to practices such as prayer, meditation, reflection, and worship. But these doctors also consistently refer to their patients' relationships with others, their sense of connectedness to the world outside of themselves. As medical practitioners they experience the presence of other people, and the support of community, to be sustaining to their patients' inner lives as well as to their physical health. And the practice of integrative medicine, they insist, is not about inserting spirituality into the doctor-patient relationship where it is not appropriate and not wanted. It is more about acknowledging that the experiences of illness and healing have always involved more than what science alone can address.

Mehmet Oz is a rigorous clinician as well a wonderful story-teller. Most of all, he is an engaging human being, immersed in the practicalities of the world around him while remaining curious and adventurous about larger patterns of meaning. And perhaps that is the definition of "integrative" living that we are grappling towards in so many disciplines. We long to bring the spiritual aspect of life constructively into play with the rest of our experiences, disciplines, and accomplishments. This is happening life by life, in creative and intellectually vigorous ways, in the most unlikely places.

*

Heart and Soul

KRISTA TIPPETT, host
MEHMET OZ, cardiovascular surgeon and author

*

The word "healing" means "to make whole." But historically, in a field like cardiology, Western medicine has taken a divided view of human health. It has stressed medical treatment of biological ailments.

As director of the Cardiovascular Institute at Columbia University Medical Center, Mehmet Oz has innovated tools and techniques, including the use of robotics that are revolutionizing the field of cardiac surgery. At the same time, as a surgeon at NewYork–Presbyterian Hospital, he's introduced mind- and energy-oriented therapies like meditation, reflexology, and massage into the operating theater and recovery room. Such therapies are sometimes referred to as alternative medicine, although many are ancient and established in Eastern cultures. The combination of alternative and Western approaches is known as integrative medicine.

But Mehmet Oz calls it global medicine. And he has a good vantage point from which to consider the convergence of old and new, East and West. He is impeccably credentialed in the best

Western schools of medical science. He grew up spending summers and holidays in the native Turkey of his parents. His father's family was devoutly Muslim, from a region known as the "Qur'an Belt" near the birthplace of Sufi mysticism and the Whirling Dervishes. His mother's affluent family came from the more secular, urban culture of Istanbul. Mehmet Oz went into medicine, he says, in part to better understand himself.

Mehmet Oz says his chief desire as a physician is to promote health in his patients and not just the absence of disease. This motivation has turned him into something of a medical explorer. And as he describes in his book *Healing from the Heart*, Oz first discovered the intense field of cardiovascular surgery to be thrilling territory for such exploration.

Oz: As you go through the process of training to be a physician, there are these "Eureka" moments, these "aha" moments that occur, particularly in the early years of medical school, where you realize some insight into existence that you didn't expect. All of a sudden, it smacks you upside your head. The heart did that to me. I remember the first time I saw this incredibly powerful organ twisting and turning in the chest cavity of an individual whose life was threatened from its failure.

The heart doesn't empty blood like a balloon letting out air. That's a very bland view of how the heart functions. It's much more elegant than that. It twists the blood out the way you would wring water from a towel. You watch this muscle twisting and turning. It looked like a cobra being tamed by the physician who was managing it.

When I saw this organ, I realized why it plays such an important role in our poetry, why it dominates our religion, why we associate the soul and love with a muscle. And I've dedicated my life to trying to figure out what that allure is and, in particular, how to help folks who are challenged with this illness.

TIPPETT: As you say in what you write, you had a very traditional, respectable American medical education, right? You went to Harvard Medical School?

OZ: I went to Harvard College and actually played football there of all things.

TIPPETT: Oh, okay.

OZ: And then from there, crazy as it sounds, I went to a joint MD/MBA program at University of Pennsylvania and Wharton Business School.

TIPPETT: Okay.

OZ: In the traditional medical training, you're told early on to pretend that the mind and the body are not connected. That you can take the organs as solitary entities—the heart, the kidneys, the liver, the pancreas, the brain—and study them by themselves. And that process is very effective for teaching people a science-based, organ-based approach to medicine.

TIPPETT: Was there a time, maybe when you were first a student, when that approach seemed sufficient to you?

OZ: Oh, it seemed not only sufficient when I was training. It was the idyllic existence, because you could really learn it. I mean, how wonderful it is to really think you know everything there is to know about the heart and the lungs and the kidneys. You get to that point of arrogance usually in your third year of medical school. You have spent two years doing nothing but studying. And it wasn't an onerous task. You actually enjoyed learning about how the body worked. You'd dream about how the body worked. And then, you're faced with the reality of dealing with people. And they don't read the same books you read. They have real problems that are different from the ones that you've been studying, because they deal with the interaction of these different organ systems. And you're forced to come to this reality. Actually, there's a story that is in *Healing from the Heart* about a Jehovah's Witness.

TIPPETT: Tell that story. That was at the end of your residency, right?

OZ: This was toward the end of my residency. Now remember, just to put this in context, you finish your medical school training and then they start calling you Dr. Oz. And you keep looking around for who that person is.

TIPPETT: Right.

OZ: And it takes about a year for it to sink in that you actually are the guy they're calling for. By the time you've gotten to your third or fourth year of surgical training, you're actually starting to become the team leader. And there was a Jehovah's Witness who was brought into the emergency room, having a bleeding ulcer. That's a problem we actually do a pretty good job dealing with these days.

But she was a smallish woman. And by the time she'd come to see us, she had lost almost all of her blood. So the solution is pretty obvious. You rush her to the operating room, and fix the bleeding ulcer by putting a suture in it. But you have to give her blood in order to have something to carry the oxygen around the body to keep her going. And the family, when I came in to talk to her, said that they didn't think she'd want the blood. I said,

"Well, that's good and all. But you realize we're not kidding around here. She's going to die if she doesn't get this blood."

So I rushed her off to the operating room. And after I had given the patient's family and her a pep talk about the fact that we needed to get the blood into her, she had become unconscious. So while she was off there, I made this last plea to the family. I said, "I'm going to do this surgery. And I'll be back to get your permission. You need to sign these forms, so I can give the blood." I went off and did the operation. By now, her blood count, hematocrit, was about four. Healthy animals start dying at a blood count of nine. She was at four and she should already have died. There was already evidence of her heart and other organs failing because they didn't have enough blood in them.

So I came out to get the permission from the family, and I was horrified to find that they were unanimous in their decision not to do this. They were condemning their mother and grandmother to death. I was flabbergasted. And only then did I really have the epiphany. They weren't telling me that they didn't believe me. They weren't telling me that they didn't love their grandmother or mother. What they were telling me is there was a deeper love, a deeper belief that transcended what I was telling them and by which they were living their lives. And that no matter how logical it seemed that they should get the blood, they didn't want the blood.

Well, as the story turns out, the woman who was going to die that evening hung out for another day, and then another day, and then another day, and she finally went home. And she never did get that blood. And although I would never recommend in the future for someone not to get the blood, it was, to me, a very revealing experience. I began to recognize that as dogmatic as I thought I could be with my knowledge base, there were certain elements of the healing process I could not capture. And even if I was right in the science, I could be wrong in the spirit.

TIPPETT: Did her recovery really defy what you had been learning all those years in medical school?

OZ: Her recovery made no sense at all. And I don't want to get into the issue of why she recovered, because there are so many hypotheses you could offer for that. But without any question, she was the first in a long series of patients. Because, you know, once you realize this is happening around you, you start paying attention a little differently. You start picking up subtle clues from patients, who may not be willing at first to share their spiritual burden with you. But now that you've expressed interest, they're willing to do that. And that, for me, became a wonderful trip, especially as I began to specialize in heart surgery, in particular with some of the sickest types of heart illness with heart transplantation and mechanical heart devices. Here are people whose hearts have rejected them. In fact, they're living a civil war.

TIPPETT: Their hearts have rejected them?

OZ: Their hearts have quit on them. Exactly. So they have to live their lives realizing that at least one of their organs doesn't think they're worthy of living. This is, by the way, how many of these folks internalize this process. And when you realize that, you begin to deal head-on with the guilt, the shame, the frustration, the anger that these folks bring to you when they need to get a new heart or they are dying of heart disease. You then get a much more robust view of the role some of these alternative and spiritual modalities may provide your patients.

TIPPETT: One thing that strikes me is that you are, and it seems that you always have been, working, as you say, at the cutting edge. You're working in extreme cases. You're working with the best new technology. And in particular, maybe you can explain this a little bit, this LVAD technology. You're working with people in that stage before they get a transplant. I'm curious if you would also say that working on the frontiers of what technology can do leads you, in some way, also to look at other kinds of therapies.

OZ: The reality is that if you're dealing with heart failure, you can say to yourself, "If only I could make a mechanical pump to keep this dying patient in front of me alive, then we'll solve all of humanity's problems," I'm being a bit sarcastic, but that's the simplistic mindset that certainly I wandered into this field with.

TIPPETT: And you also have patents for tools you've developed. So you're doing that also, aren't you?

OZ: Exactly. I spent a lot of time trying to figure this out with the hope and the belief, the passionate belief, that if I could make some of these devices work, we could actually get folks to not die of heart disease.

Well, guess what? I'll tell you this story because it's actually reflective of this. I had a gentleman, a very religious man, and religious as defined by the fact that he was a churchgoing fellow, who spoke frequently of the power of his faith. I learned this later on about him. But he used to drive the sand machines during the snowstorms in upstate New York. And he had a massive heart attack and basically dropped dead while working. He was rushed by a helicopter to our area and eventually to our institution, where I realized that his heart had died. And the only hope to keep him alive was to put a mechanical device in him, a so-called LVAD, left ventricular assist device. These devices are pumps that act as a piggyback support system because the heart itself can't pump blood anymore. The surgery went wonderfully well. He recovered from his operation. I had never met him, remember, because he was unconscious when he came to us. And the first time he met me, he told me he wanted to kill me and then kill himself to follow.

Now, here I am giving myself a rotator cuff injury, congratulating myself, patting myself on the back. And he's telling me that he doesn't want to live anymore. In talking to his wife, I learned that he had lived under the assumption that he would always play a valuable role in the world. And when he no longer could contribute to the world, he would be allowed the dignity to die. Here I had taken that dignity from him. I had forced him now to live as what he perceived of as a cripple with no value, no use.

So the way we dealt with this problem, with the help of his wife and his pastor, was to get him involved as an evangelical force within his church. And this gentleman, who subsequently got heart-transplanted, now actually provides ministerial services for Hells Angels motorcycle gangs. It was, for me, a wonderful example of the fact that people crave a use in life, and if you take that from them, you have to try to replace it in another context.

TIPPETT: And that that is a part of healing?

OZ: Ultimately, the healing process transcends the replacement of the organ and moves into the spirit. And that's where disconnects happen. When you finally figure out that you've got the best technology available, when you've finally climbed the last technology mountain and the patient still doesn't feel well, you've got to look elsewhere. That's when we start looking in areas where we're much less comfortable, like spirituality and alternative therapies that bridge cultures of healing beyond this country's borders.

TIPPETT: You do use many different, what we call alternative therapies or traditional therapies. Talk to me about some of these therapies, how you've come to them and why they've come to seem important to you and how you experience them to be working.

Oz: In many cases the alternative therapies were brought to me by folks outside of medicine. Within the institution that I work in, in New York–Presbyterian Hospital, I found that there were patients who came to us from all parts of the globe who had their own healing traditions that had been effective for them in the past. They wanted to use those, but they kept feeling that we didn't want that to happen. They would abdicate all responsibility for their care once they walked into our hallowed hallways. And so we tried to change that. We tried to give them the confidence to play an active role in their own recovery process by letting them use their own healing traditions. And that's how I actually learned about many of these alternative therapies.

TIPPETT: So is it your sense that in other cultures, where what we call traditional therapies are the primary therapies, health care is more interactive? And are patients in the West more passive?

Oz: I feel strongly that in the West we have come to believe that medicine offers all the solutions and so we no longer play the proactive role we should be playing. Take Turkey as an example. You would never leave a patient in the hospital there unless you had a relative with them. In fact, the nurse gives you the pills to give the patient. You change the bedpan. You make them feel comfortable. You fluff up their pillow. In the United States, we have visiting hours. No one can see the patient. We block them out. We create barriers to the family and the loved ones playing a healing role for the individual who's sick. These are the kinds of disconnects that we have created because we've had so much trust in science. And please, I have a lot of confidence in science. In no way do I wish to bash the field that I'm so proud of, of medicine.

It's just that if we're truly going to achieve maximum healing, maximum impact, we ought to take any tool that's at our disposal. That includes nonscientific approaches, as long as we have evidence that they don't hurt the patients. And that's really what I'm pulling for.

In his surgical practice at New York–Presbyterian Hospital, Mehmet Oz has recommended and integrated complementary forms of patient care such as hypnosis, yoga, and work with the body's energy fields as understood by Tibetan Buddhism. Oz says that as he assesses such practices he keeps his mind both open and discerning. He has to be satisfied by the same standards with which he assesses Western techniques—that is, whether they actually work for his patients. In order to explore and document the complementarity of Western and alternative treatments, including the role of spirituality in healing, Oz cofounded the Columbia Integrative Medicine Program. This is part of a growing movement of such centers at leading hospitals and medical schools across the country.

✳

TIPPETT: Talk to me about some of the tools that you treasure the most. You write a lot about hypnosis.

OZ: Hypnosis is a therapy that is, I don't think, even that unconventional anymore. But we have studied it in numerous different settings. There are many other individuals across the country who've also done work along these lines to demonstrate that hypnosis can play a role in ailments as varied as hypertension to the chance of having pain after a procedure. So I divide these alternative therapies into two basic camps. There are the alternative therapies where you put something in your mouth; herbs, vitamins, and all those things. Let's leave those to the side because they really get into the science and medicine of what we're doing.

TIPPETT: Okay. And even homeopathy, would that be in that category?

OZ: I would put homeopathy in that group as well, though, of course, homeopathy works in a very different way.

TIPPETT: Yeah.

OZ: And then there are the therapies where your mind plays a role. And what we're really trying to do is to figure out how to get your mind—and perhaps elements of your mind that we don't understand—working with you. Let's take the big area of energy. Whether energy exists or not at the macro level, at the level of the human being, is a difficult thing to tell. But we define life at the level of the cell by whether or not you have an energy level in the cell that's different from the energy level outside the cell. That's what life is. So if you aggregate those cells together into an organ, the heart, and you put those organs together into a body, the human, why would we think that we wouldn't have energy that's measurable and could be affected to make you feel better?

In fact, why would we not think that disturbances of that energy might cause some of the ailments that we cannot today put a name on? That's why I think therapies like acupuncture and Tai Chi and acupressure and even the use of some of these medicinal treatments like homeopathy, which may affect energy levels, could actually be an important advance for us in medicine. If nothing else, it widens the vista of opportunities that we have in the healing arena. The big challenge is, it is very difficult for folks to invest the resources to truly study these modalities. And because they are underfunded, it is often impossible to envision a mechanism to truly "prove" that a therapy can be effective.

TIPPETT: Let's say something like acupuncture. My understanding is that a Chinese physician or healer and a Western physician, while they might share a sense of basic human anatomy, they have very different paradigms for understanding how the body works. And maybe it comes back to this idea of energy. You can explain this better than I can. But is it your experience that these different paradigms are not in contradiction but can be brought together in one medical practice? Or is there anything you're grappling with that is simply asking you to divide your mind in two and say that these are two worldviews that don't match?

OZ: There are definitely situations where the therapy I would term as alternative would not work well. An example might be homeopathy, because in homeopathy you're assuming that small amounts of a product can influence the way the body responds. And because we can't predict what that response is, it's hard to use that in conjunction with a beta-blocker or Lipitor.

TIPPETT: Okay.

Oz: That stated, there are many, many other areas, the vast majority, where I can see them working quite effectively together. Take chemotherapy, for example, which would be used against a particular cancer. It causes symptoms—nausea, vomiting, hair loss, and the like—which could be ameliorated by the use of alternative therapies. In addition, we could use green teas and a variety of mind-body elements, including the use of music and guided imagery, to impact tumor growth rates. And from my perspective, what's really happened is the globalization of medicine. Think about this, Krista, for a second. We have global media. Your show can be watched anywhere or listened to anywhere. We have global banking and finance. We have global entertainment.

We don't have global medicine. And that's because medicine is a remarkably provincial process. The doctors come from their local culture, they have the same biases as their mothers gave them. And so they go out and start practicing using therapies that they think work and ignoring ones that may work but that they don't think work. So alternative medicine has really become the globalization of medicine. It is incorporating healing traditions from other parts of the world. And in sort of carrying this to the ultimate extreme, we just finished a nice study with Mitch Krucoff and the folks from Duke looking at the role of prayer in healing. And this trial, which was called the MANTRA trial, was a randomized trial, but we actually got groups to pray for the patients from Tibet, from France. We had Baptists, we had Protestants, we had Catholics, we had groups of prayers from all the major religions in order to assess whether prayer might play a role in the recovery of folks who had heart problems. And this is the kind of globalization process that I suspect will grow over the next few years.

TIPPETT: Now, you mentioned the Randolph Byrd study in your book, which is one of the most famous studies of prayer and healing. But it then became very controversial. There's a lot of skepticism and controversy around all of these prayer studies. So I am curious about where you come out on prayer as a part of healing.

Oz: Well, at the outset, I should say that I also entered into the study of prayer with some reluctance, in part because I had felt that maybe we shouldn't be meddling with prayer. Maybe that was too personal, and who are we to start trying to examine something as potentially powerful and also misleading as prayer? I was comforted by a pastor who told me that folks a lot smarter than I had tried to destroy religion before, and I should feel comfortable doing this research. So we began to go after it in a fairly substantial way. And the Byrd study, which demonstrated a seeming benefit of prayer in folks who were in an ICU . . .

TIPPETT: Who were prayed for, right?

Oz: . . . they were prayed for, and the people who got prayed for did better. It is a trial that is one of several that have looked at this topic, and all have been faulted because they weren't large enough and they weren't randomized the way they perhaps could have been. So we decided to put together this large 750-patient trial. But of course you run into problems with endpoints and what were the biases of the patients. For example, 90 percent of the people in the trial thought they were getting prayed for already.

TIPPETT: By people they knew?

Oz: By people they knew. So it becomes difficult to tease out if your prayer's doing it or their prayer's doing it. But we did wander upon some interesting observations. And here's one that may blow your mind, so to speak. There was a trial that had been done by a group from Korea looking at the role of double prayer. In other words, not just a prayer group for your patient, but a group praying for the group praying for your patient. This seemed far-fetched to me. I had no idea. And the reason they had done it was because they were in Korea and they were a Christian hospital, so they wanted people praying from the States and they wanted to power it up a little bit. Again, this is perhaps a very simplistic view of how prayer works, but nevertheless they had seen some benefits in the fertility rates in their study. So we did that at the end of our trial.

TIPPETT: In the MANTRA study?

Oz: In the MANTRA study. And we saw some intriguing findings. Again, it was only in the last part of the trial, but we saw changes that were enticing to us and have prompted us to want to do a follow up study looking at that particular tool and the role that it may play. But people get fixated on the subtleties of the studies. At the end of the day when you do studies on religion, you deal with religious biases. If in your heart you don't think religion will play a role, then you will find the data sets that support that. And if in your heart you think that prayer will work, then you're going to find information that supports that view. And the smarter you are, the better you are at finding data to support your biases. This is the fundamental disconnect we have as rational human beings trying to deal with faith. And it is a challenge that I face day in and day out with folks who are coming to grips with what meaning their ailment has for them.

I'm reminded of a story that happened recently of two fathers who came to see me, both of whom had heart disease. The first father came with his wife and told me that he didn't really care if he survived this heart surgery. And I said, "Well, that's not a good place to start off the discussion." And I started to probe a little bit into why he didn't care if he survived.

It turns out that his young boy, a sixteen-year-old kid, had died in a case of mistaken identity. This child had been his dream-child-come-true. He had had such a good time with the kid. He was a wonderful kid. And when they lost their child, they had become despondent. And the heart disease that occurred afterward to this gentleman was almost a blessing because it might provide him an excuse to exit this planet.

So I said, "We're going to talk about this," and I sent him home. I just didn't even know how to begin to address the grief he obviously felt from losing his son. But I knew that he could not enter any kind of a life-threatening process like heart surgery, much less life, with that kind of attitude.

That same week, a father came in to talk to me. He walked in and the first thing he said was, "Doctor, I have blockage in my arteries. You have to operate, and I have to live." I said, "Well, of course, you want to live." He said, "No, no, no. I don't mean to interrupt. I have to live." This intrigued me and I said, "Why?" He said, "I've got a retarded child at home. He's profoundly debilitated. I have to change his diapers. I do everything for him. If something happens to me, there will be no one there to take care of him. I have to live."

Now put these together. The second father never enjoyed having a game of catch with his son. He never went to the movies with his son. He never watched his son play any musical instrument. He never had the kinds of blessings the first child had. And yet, he saw an element of grace in the existence that he had with this sick child that drove him to want to live. And when I shared this story with the first father, it changed his outlook as well. At the end of the day, being ill is an opportunity for us to learn more about why we're here. Some folks climb mountains, others get to have heart surgery, I'll often tell my patients.

TIPPETT: So how do all of your experiences as a doctor change your definition of what "quality of life" means?

OZ: Well, quality of life has changed a lot for me as I've witnessed patients. For me, it was initially just life. You know, being alive was quality of life. And it is true that if you're not alive, there is not much quality. But staying alive by itself is not the only goal. And we as a society have to mature our views of death and dying in order to cope with the reality that we have science now that can do more than we want it to do. Quality of life has become a dominant element of my discussions with patients.

I've had older Americans come to my office and tell me that although they are perfectly physically able to have surgery, they didn't have anything to live for. All their loved ones had passed along. Their families had gone their different ways. They were pretty much just biding their time, waiting. So why would they bother having life-threatening surgery that would just prolong their existence when they had had a great life? And by the way, they're not depressed. They've had a great life and they've done it. They're ready. And that is a conversation that would have troubled me much more when I was younger. When someone tells me that now, and they have good reason to say what they're saying, I'm accepting of that.

TIPPETT: And then you would not perform the surgery?

OZ: It's not even a matter of performing the surgery. As a physician, you have a precious covenant with your patient. And because they generally trust you, you can talk them into things. So it's not a matter of whether I would do it or not, it's whether I would try to talk them into something that maybe I wouldn't talk myself into when I was in their shoes. And although I just turned forty-four—so I can't truly identify with an eighty-eight-year-old patient, twice my age, who might feel this way—I begin to see the wisdom, at least, in that discussion. After all, if you don't have a good reason for your heart to keep beating, it usually won't. And some of these folks have thought that process through better than I have.

TIPPETT: I want to come back a little bit to this idea of prayer. I would like to know, through the study you took part in and through using this technique in your work as a heart surgeon, how have you come to think about what the value is of prayer, what's happening in that, how that can be legitimately integrated into medical care?

Oz: Well, we never prayed in the MANTRA trial. We never asked the prayers to pray for the patient to survive. We asked them to pray that "Thy will be done." We asked them to pray for what was best for the patient to happen. So maybe if you're eighty-five years old and you have metastatic cancer and you've got no one left in the world, maybe the answer to the prayer is to let you go grace-fully from a heart attack, which is, after all, not the worst way to go. It's painless and it's quick. So we do have to be cautious, as the saying goes, for what we wish for because it might come true.

But I do think the opposite approach would be to ignore the potential power of prayer. Again, I do want to be cautious. When I speak of prayer, I'm not even talking particularly of the orga-nized religion behind the prayer. It's really the role of spirit and whether or not there's an energy behind this spirit that we can tap into and take advantage of, an energy that is spoken of in most religions, and that we generally completely ignore in West-ern medicine because we can't measure it. It would be, I think, an abdication of my responsibility as a healer to not at least look into those opportunities.

I've always have been intrigued by this. You called me Dr. Oz earlier. Now, the word "doctor" comes from the Latin root for "teacher." But, you would also say, I went to medical school. Well, "medicine" means "healer." And "physician" comes from the Greek for "physics" or "science." So even in the way that you call me what I am, you're describing me as a teacher, a healer, and a scientist. So I need to be able to wear three hats on top of one another or at least shift gears between the three opportunities. And science, unfortunately, meets a roadblock once in a while. As we wait for that paradigm-shifting understanding or insight that will allow us to go to the next level with science, which I'm confident we will do, we sometimes have to allow elements of faith or belief or insight or intuition.

For example—and this is perhaps a little bit off the topic—what gave Einstein the idea that there were particles or waves in physics? Is it possible that he was colored at all by looking at Impressionist paintings that had been done for the past thirty years when he was formulating his ideas, which created light from dots? As in that example, perhaps art colored the thinking of—if not Einstein, then other physicists of the time. Medicine and physicians, we have an understanding of energy. We have a digital world. We have insights into technologies that we haven't yet applied in the context of the human body that we will probably one day, in this next generation, gain insights to.

TIPPETT: So when I read your story and read about you, one thing that jumps out at me that's rather simple, but very profound in its effect, is that while you are a highly trained, highly skilled doctor, you're also very open to seeing what's happening with your patients—and even experiencing the birth of your own children. And you're always questioning the limits of medicine, and then reaching out for other resources—in your case, alternative treatments. And I wonder if maybe—you said you're forty-four now—do you think this is a generational shift? Do you think that more doctors your age are simply more open to the complexity of the whole experience of healing and health?

Oz: I think there are many more opportunities for younger physicians to get that exposure. In part because the generation before us was still striving to figure out the basics of how to keep folks alive using science.

In 1955, you would not have had heart surgery because we couldn't do it. In 2005, I can do two operations in the morning and be on a radio show in the afternoon. It's a completely different world. In 1955, my main goal would be to save that kid's life using new insights in science that even two years earlier didn't exist. In 2005, I know I can save that child's life, but I know that there are elements of depression and disconnect that might occur in the postoperative period. And I know that even more important than the hole that I fixed, there are other issues that will challenge that child that I need to get addressed if I'm doing my job as a healer.

So the game has gotten more complicated. And because we have the honor of standing on the shoulders of our forebears, at least in medicine, we can see farther. I can see the mountain in the distance. I can dream about things that they didn't have the luxury of dreaming of because patients were dying in front of them for reasons that they thought they could easily fix. People don't die in front of us today for easily fixable reasons, and that pushes us to look a little farther for true healing.

TIPPETT: But what's ironic and so interesting is that some of the places you're looking are ancient traditions that previous generations of doctors would have considered to be very simple, would think that the West had outgrown. Right? I mean, acupuncture or . . .

Oz: Absolutely. But that is the globalization of medicine. And as we explore beyond the borders that have traditionally limited us, it takes us to places where we're not too comfortable. But that's what it's all about. In a way, for me, life is about being comfortable with being uncomfortable. It's about taking yourself and the people that trust you on a life journey, because that's what health is all about. And we all have our own individual health parade through life. It's a serpentine path that takes us to places we didn't expect, but that's part of our life experience. Our job may be to incorporate approaches that we never could have envisioned playing a role in recovery. But that now, because we have the luxury of looking a little farther, we can identify.

TIPPETT: When we first started speaking, you described going into medicine and wanting to make the world a better place. And it sounds to me like being a doctor and working at the cutting edge of science in fact has made you a more spiritual person. Is that right? Is that true?

Oz: There's no question that I've become more spiritual because of the practice of medicine, particularly because I wandered into a field that was high tech. And so the illusion that I could find salvation through science alone was no longer present.

TIPPETT: Can you say something about how your particular spiritual sensibility or practice has been concretely shaped by your experiences as a doctor?

Oz: Well, for one, as I look at how my spirituality has changed, I've become more comfortable re-exploring spirit. There was a time in my life where I spent a lot of time thinking only on this topic, and it was actually during my college years when I was not atypically trying to just figure what the heck was going on so I could get on with my life. And, as many folks do, I got on with my life and for fifteen years or so didn't think much about religion beyond the necessary elements of making sure the kids, you know, went to Sunday school or that we dealt with the religious holidays. But as I've grown more and more attuned to what my patients are asking for, I've become more insightful to my own needs.

And I do want to correct one thing you said that was kind about me. You said that I went into medicine to make the world a better place. And although, without being falsely modest, that was truly a driving force for me, there was clearly a narcissistic element to this. I really wanted to study me. I wanted to know what was going on. I wanted to be an explorer, and I wanted to know about why we are here and what we are doing here. I thought medicine would take me there. And it has, but not all the way. To continue the journey, I have to go beyond where science, in its traditional context, would take me. I have to look for clues to what the next steps may be. And spirituality helps me along that path quite a bit.

In fact, a lot of my personal interest in yoga comes from a recognition that I can reach a Zen experience, a blissful existence, if I can get my body and my mind calm together. Yoga does that for me as well as any other element. I appreciate hymns, chanting, much more today than I did when I was a schoolchild because I see in that a sense of peace and emptiness that frees me. These are insights that I think you have to be a bit more seasoned—at least I felt I had to be a bit more seasoned—to appreciate. And without the insights that medicine has provided and my teachers or the patients have provided me, I wouldn't have wandered upon it.

TIPPETT: People who are close to death often experience a sense of a reality—of another level of reality. As a surgeon who is sometimes with people in those moments when they're hovering between life and death, do you experience something palpably?

Oz: I don't normally experience the near-death elements, in part because I'm pretty busy trying to prevent the death. But there is no question that you sense a deep-seated loss when a patient dies. And it doesn't go away. You can hide it and bandage it better as you get more experience dealing with death. But when someone leaves and you didn't want them to leave or you don't think they wanted to leave, the sense of loss is deep. It's a coldness that's inside of you, and it takes another person to get rid of it, either the family member of the patient or your own family, in my case, where I go for recharging. But that is a very draining experience, and it's something that I suspect one day we'll be able to put numbers on and measure and quantify. But for today, I would just call it sadness, a cold sadness.

It's something nontangible, unmeasurable. If I was using a Harry Potter analogy, I'd say there was one of those goblins that had come in and stolen my very chi, my very essence.

Tippett: You have a lot of lovely quotations in your book, Sufi quotations, also Maimonides, all kinds of people. But there's one in the body of what you've written, it's William Blake. And there's just something in the way you put it into the text that made me think it's really meaningful for you. I want to read it and ask what this means for you as a person and as a doctor. Blake wrote, "To see a world in a grain of sand and a heaven in a wild flower, hold infinity in the palm of your hand and eternity in an hour."

Oz: William Blake was actually Swedenborgian. Swedenborg is the Swedish philosopher whose writings resulted in a Protestant sect after his name which is based in Bryn Athyn, Pennsylvania, and it's my wife's religion.

I was particularly attracted to the writings of Swedenborg because they provided a clarity that I found lacking in many other traditions. And William Blake's quote so beautifully identifies that. What he's really talking about is this concept of complementarity, a term that was coined actually by Niels Bohr, the famous physicist in the 1920s. Complementarity was a term that meant that you could have two mutually exclusive answers to a problem and they could both be right. Now how could that be? Well, in physics, it was wave theory and particle theory. It was a thought that energy could be both in a bolus and in a wave. Why? Because it didn't actually ever exist in either form. It was a tendency to exist in a particular location that defined it. And once you got past your concrete thought processes about what energy was, you could actually come to peace with this complementarity of reality. William Blake is talking about the same thing. How can the world be in a grain of sand? How can infinity be in a second? How if these are mutually exclusive possibilities? It challenges your basic underlying understanding of what reality really is. And when you move past a physical understanding of reality and start to acknowledge a more spiritual foundation for what reality truly is, you begin to realize that we live in a world where 99 percent is pretend and 1 percent is real. And what we're striving for as human beings is that unmodulated experience, that unmitigated exposure to the 1 percent of reality. And that's where medicine has taken me, and that's where patients who are struggling to survive are going.

TIPPETT: I certainly hear the analogies in this idea of complementarity and what you are exploring and experimenting with in medicine, which might seem to some to be two very different worldviews of Western medicine and traditional approaches to medicine. You've also observed that traditional medicine does make room for a nonphysical aspect to the human being, to energies that can be involved in healing in the way that Western medicine doesn't. There is this acknowledgment of a reality of transcendence in these lines of Blake as well.

Oz: Yeah, I think Blake highlighted that beautifully in his poetry. It's evident in many of the stories that we face in our lives, but we have to open our eyes and our ears to hear and see them. And that's often where our shortcoming is. That's where, crazy as it sounds, being ill offers you a growth opportunity because you're much more willing to pay attention to subtle things if you have the threat of that experience being taken away from you.

4

＊

Creation as an Unfolding Reality

"EVOLUTION AND WONDER"

Charles Darwin published *The Origin of Species* in 1859. We've come to imagine him as a godless naturalist and to see the publication of this book as a dramatic moment in history, one that created an instantaneous rift between science and religion. These assumptions fuel some of our most intractable cultural debates.

In my conversation with the biographer James Moore, we reject those debates. We explore the world in which Darwin formulated his ideas. We read from his varied writings. We ask what Darwin himself believed. Did he find in his observations of the natural world a rejection of God and of creation? How might he speak to our present struggles over his legacy?

As it turns out, Darwin was grounded in the distinctly reverent Judeo-Christian philosophy of Western science up to that point in history, a view of the world encapsulated in a quote of Francis Bacon that he put opposite the title page of *The Origin of Species*:

Let no man . . . think or maintain that a man can search too far or be too well studied in the book of God's word, or in the book of God's works . . . but rather let men endeavour an endless progress or proficience in both.

Darwin, as we learn from James Moore, was agonizingly aware of the fixed worldview that his theory of transmutation—the original term for evolution—would unsettle. The people of Darwin's time believed that every condition of plant, animal, and man was static and eternal, brought into being all at once at the beginning of time.

They estimated that to have been six thousand years earlier. But *The Origin of Species* was not the first classic scientific text to break from such beliefs. It was, rather, the last to fully engage them. Darwin waited two decades before he published. His observations and conclusions were painstakingly belabored. He anticipated religious questions and objections at every turn and responded carefully to them. Darwin's theory of natural selection was born, James Moore asserts, of "theological humility." This insight alone would place our culture's contentious battles over Darwin on a different footing.

My own suppositions have been radically changed by this discussion. I'm reminded of the conversations I had on Albert Einstein. Einstein did not reject the idea of a force or "mind" behind the universe. But he saw that expressed in natural laws that could be discerned and described.

In a similar way, Darwin saw creation as an unfolding reality. Once set in motion, as he saw it, the laws of nature sustained a self-organizing progression driven by the needs and struggles of every aspect of creation itself. The word "reverence" would not be too strong to describe the attitude with which Darwin approached all he saw in the natural world. There is a great intellectual and spiritual passion and a touching sense of wonder evident in his writings, from his private notebooks and correspondence to the *Beagle* diary and *The Origin of Species*.

For me, this view from within Darwin's life and times opens up fascinating new ways to ponder not the rift but the possibilities for exchange between science and theology. He used the biblically evocative analogy of a "tree of life" to illustrate his theory of species sprouting as branches from the same trunk, some flourishing and others withering and falling to nourish the ground in which the whole is sustained. His vision of all of life netted together is profoundly consonant with what we are learning now in environmental sciences as well as in genetics.

In describing a creation that organized itself, incorporating chaos and change into survival and progress, Darwin did not challenge the idea of God as the source of all being. But he did reject the idea of a God minutely implicated in every flaw and injustice and catastrophe.

As James Moore puts it, Darwin forced human beings to look at the inherent struggle of natural life head-on, not as we wish it to be, but as it is in all its complexity and brutality and mystery. This is most difficult for human beings, perhaps, in times of great change and turmoil such as ours. Indeed Moore and I trace the fact that the greatest resistance to Darwin's ideas has appeared in other cultural moments of flux and global danger. But Moore tells his students who believe they must choose between belief in a creator and the science of Darwin simply to read *The Origin of Species*. There is much in Darwin's thought that would ennoble as well as ground a religious view of life and of God. I'll end with that book's final lines, which are rich with wonder:

[F]rom the war of nature, from famine and death, the most exalted object which we are capable of conceiving, namely, the production of the higher animals directly follows. There is grandeur in this view of life, with its several powers, having been originally breathed by the creator into a few forms or into one; and that, whilst this planet has gone cycling on according to the fixed law of gravity, from so simple a beginning endless forms most beautiful and most wonderful have been, and are being, evolved.

Evolution and Wonder

KRISTA TIPPETT, host
JAMES MOORE, biographer of Charles Darwin

✻

Charles Darwin published *The Origin of Species* in 1859. He was the son and grandson of physicians, a gentleman in early nineteenth-century Britain. He grew up in the world of Jane Austen's novels, a world of manners, politeness, and of a rigid class structure.

This social structure was held to be divinely ordained, like every condition of plant and animal, fixed and static and eternal. The Protestant Reformation of the sixteenth century had brought biblical certainties to laypeople in their own language, and they read the story of creation more literally than the classic theologians had.

Though he was a passionate amateur naturalist, the young Darwin was headed for a career in the church. But first, at the age of twenty-two, he seized a chance at adventure, a place on the near-five-year scientific journey of HMS *Beagle*. This took Darwin across the globe and to the southernmost tip of South America. There he observed a vast and vigorous spectrum of life that filled him with amazement and with questions:

How have all these exquisite adaptations of one part of the organization to another part, and to the conditions of life, and of one distinct organic being to another being, been perfected? We see these beautiful co-adaptations most plainly in the woodpecker and mistletoe, and only a little less plainly in the humblest parasite which clings to the hairs of a quadruped or feathers of a bird, in the structure of a beetle which dives through the water, in the plumed seed which is wafted by the gentlest breeze. In short, we see beautiful adaptations everywhere.

Our guide to understanding Darwin is his biographer, James Moore, a Cambridge research scholar who's studied and written about Darwin for three decades. Moore grew up in a fundamentalist home in Chicago, where he learned to think of Charles Darwin as an enemy of God. Darwin had feared that his ideas would be characterized in this way. After he returned to England from South America, he waited nearly twenty years to publish his theory of the origin of species. He once wrote to a friend that this felt like confessing a murder. I asked James Moore what Darwin meant by that.

MOORE: We have to look at the mood at that time and in all of the years Darwin was being educated. God was in his heaven, all was right with the world. At least in England, people knew their places. Things were changing, but it was widely believed that both society and the natural world were held stable, fixed, by God's will. And this world was justly and correctly administered by God's agents on Earth, his priests. Species did not change spontaneously and naturally, because nothing in this world happened purely naturally and spontaneously. God was in charge. When Darwin confessed to murder, he was saying that nature is self-developing. God, according to Darwin, had established laws by which matter moves itself and changes into new forms we call species. Darwin was not denying God's existence. The murder was not the murder of God.

TIPPETT: I think that at that time, in Victorian Britain, the whole field of biology was captive to creationist theology. But I don't think it had always been that way. Is that right? I mean, was it particularly true in that era?

MOORE: We have to use the word "creationist" or "creationism" very carefully. Historically, Christians and Jews and Muslims are all creationists because they believe that God brought the world into existence. A creationist was not a person historically who had any particular views on the origin of biological species, but was one who held certain theological views about the universe and about the soul.

The definition of "creationist" became narrowed in the seventeenth century and in the eighteenth century. At this time, people were discovering a great deal more about the natural world and were classifying individual species and grouping these species in larger and larger groups. And it became a matter of belief during the seventeenth and eighteenth centuries that each of these species, each of these biological species of plants and animals—tens of hundreds, thousands of species—had been individually created by God in their first pair in the Garden of Eden. And the poetry of John Milton in *Paradise Lost* gave a great deal of color to that.

*

Milton's *Paradise Lost* was among the four books Darwin took along on HMS *Beagle*. Here are some verses:

Let us make now Man in our image, Man
In our similitude, and let them rule
Over the fish and fowl of sea and air,
Beast of the field, and over all the Earth,
And every creeping thing that creeps the ground.
This said, he formed thee, Adam, thee, O Man,
Dust of the ground, and in thy nostrils breathed
The breath of life; in his own image he
Created thee, in the image of God
Express; and thou becamest a living soul.

MOORE: There's a literalism in this poetry that Christians took to be part of the explanation of the origin of biological species. So by the time Darwin is born in 1809, it is a common assumption in all churches and by all Christians that the original pair of every species had been brought into existence not so long ago by God. This was a modern belief. It was not a common belief before the seventeenth century.

TIPPETT: That's really interesting. What you're describing is, as people began to learn, as science kind of opened up and people began to learn more about the natural world, there was an attempt to fit that knowledge into the biblical stories. But the result of that was to make those more rigid than in fact they were. I think previously even theologians didn't try to make Genesis a scientific text. They read it as a theological text with a theological purpose.

MOORE: Ordinary people read the Bible with their ordinary spectacles on. The people who told them what the Bible says were very, very important. In the Protestant Reformation, those people were not to be the church dictating how you read the Bible, but the individual believer. So the Bible became an open book much more than it had been when it was translated into the vulgar language, the ordinary language of people. I believe the Catholic Church was right to this extent, that this really did open up a Pandora's box of possibilities. Because with every person becoming his or her own interpreter, there was scope for really quite extraordinary clashes about what God is telling us through this book.

And as far as the creation story is concerned, of course, we don't know what God has created without looking around us in the world. So with voyages of discovery, with intense national investigations, we began to build up a picture—people began to build up a picture of an extraordinary diversity of life on Earth. And that had to be fitted into the ordinary person's view of the Bible.

In Darwin's time, literal readings of Genesis were based on an assumption that the Earth was no more than six thousand years old. Darwin addressed this assumption directly in *The Origin of Species*:

> The belief that species were immutable productions was almost unavoidable as long as the history of the world was thought to be of short duration; and now that we have acquired some idea of the lapse of time, we are too apt to assume, without proof, that the geological record is so perfect that it would have afforded us plain evidence of the mutation of species.
>
> But the chief cause of our natural unwillingness to

admit that one species, has given birth to other and distinct species, is that we are always slow in admitting any great change of which we do not see the intermediate steps . . . The mind cannot possibly grasp the full meaning of the term of a hundred million years; it cannot add up and perceive the full effects of many slight variations, accumulated during an almost infinite number of generations.

TIPPETT: This is something that we have no historical memory of in our present culture, but it's very clear when you start reading this book that there is this painstaking care that Darwin makes with every observation of the natural world. It's almost like he's anticipating the theology that he is challenging or trying to open up. He's at this moment where religion and science were joined and then there starts to be a divide. But he's right there before that divide actually takes place, is that right?

MOORE: Darwin's understanding of nature never departed from a theological point of view. Always, I believe, until his dying day, at least half of him believed in God. He said he deserved to be called an agnostic, but he did make the point later in life that "when I wrote *The Origin of Species*, my faith in God was as strong as that of a bishop." So Darwin's many references to creation, there are over a hundred references to creation in *The Origin of Species* . . .

TIPPETT: When you really read the text, you are aware of the struggle. He is wanting to be respectful. He takes very seriously the religious and cultural assumptions that he realizes he's disturbing.

MOORE: This is what I tell my students: if you are a creationist or you're inclined to be sympathetic with what we now today call creationism, read *The Origin of Species*. Darwin wants to convince you in this book that God has established laws of nature on Earth, as in the heavens, and these laws produce the forms of life that we observe. And the principal cause of this, for Darwin, is what he calls natural selection.

TIPPETT: At the beginning of *The Origin of Species*, he puts a quote from Francis Bacon. I want to read it, and I'd like for you to explain what this was describing in terms of a way of looking at the world and why Darwin put it at the beginning of *The Origin of Species*.

Bacon wrote, "Let no man . . . think or maintain that a man can search too far or be too well studied in the book of God's word or in the book of God's works . . . but rather let man endeavour an endless progress or proficience in both."

MOORE: This is Francis Bacon, the philosopher, the statesman, writing in the seventeenth century. The two books for Francis Bacon are the "word" of God and the "works" of God: the Bible, and the works of God in nature.

TIPPETT: The works of God is everything we see around us, right? The world.

MOORE: The natural world.

TIPPETT: The natural world.

MOORE: And for Bacon, it's important that the works of God teach us how to interpret the word of God. So what we see in nature . . .

TIPPETT: Rather than the other way around, isn't it? I think if there is an attempt in our time to look at this, it's the other way around: to interpet the works of God through the word of God.

MOORE: There's been a reversal, and people have gone off on some extraordinary tangents in so doing. For example, opposing Newtonian astronomy on the grounds that the book of Genesis rules it out. So right at the front of *The Origin of Species*, Darwin has a quotation from the revered Lord Bacon, to show that the Bible and natural history should be studied together.

TIPPETT: Now, as you say, we associate Darwin's name with the split. But until then, even some of the scientists that we think of as opposed to the church—Newton, Galileo—they also were in that tradition of seeing their work, understanding the world—the created world, they might have described it—as illuminating Christian tenets in the Bible.

MOORE: Absolutely. Absolutely. It's very important to realize that in return for telling us how texts of the Bible should be interpreted, people who investigated nature, call them naturalists, were also expected to supply evidences of God's beneficence, power, and wisdom in the works of nature. So the marvelous way in which a bivalve shell is constructed, or the wonderful joint in your elbow, or the patterns of life, the beauty of butterflies, all of these things can be studied by naturalists and said to be evidence of the Creator's wisdom and beneficence.

TIPPETT: And Darwin really is in that line. That was his inheritance, in a sense.

MOORE: Darwin's starting point were these wonderful—the term was "adaptation"—the wonderful adaptations of organisms to their environment. Things seem to be made perfectly to live where they are: fish to swim, ducks to paddle, and so forth. These traditionally were evidences of the Creator's wisdom and goodness. Darwin says, "We can explain how nature produced these adaptations to environment. We can explain how the beauty of a butterfly is useful to that butterfly in pursuing its way of life. I can come up with causes for this and it's up to you to believe that God created these things through these causes or not."

Darwin evokes the works of God, the works of natural theology, the greatness of nature, at the beginning of *The Origin of Species*, because he really does believe those works in nature are beautiful and astonishing, and the adaptations of their—he's at one with the spirit of natural theology. Just read his prose in *The Origin of Species*. It exudes wonder of nature—but he can explain how it happened.

TIPPETT: I wonder if you would tell some of the stories you've told in your writing, some of the turning points for Darwin, moments during the voyage of the *Beagle*.

MOORE: Darwin sailed on HMS *Beagle* in 1831, a fairly conventional product of Cambridge University. He had been brought up in one corner of one culture in western Europe. He had never seen a person without clothes on, never seen a woman without clothes on. And suddenly he's thrust into a situation where, immediately on landing in Brazil, he sees slaves being traded. He sees people in chains and in servitude to other people. His whole family hated slavery, but now he confronted it. And it was about this time that he wandered off into the forest for the first time and sat down on a mossy log and made notes in his field notebook, and he actually uses a word from the Bible. He says, "Hosanna." He sees the palms around him, as on Palm Sunday:

In Bahia, Brazil, April 1832. Sublime devotion the preva-
lent feeling. Started early in the morning. Pleasant ride
and much enjoyed the glorious woods. Bamboos 12 inches
in circumference. Several sorts of tree ferns. Twiners en-
twining twiners. Tresses like hair. Beautiful Lepidoptera.
Silence. Hosanna.

Later he reached the southern tip of South America, Tierra
del Fuego, and here he sees what he calls real naked savages for
the first time. He sees a full-term pregnant woman with rain and
sleet dripping from her body. He hears what he describes as an-
imal-like sounds coming from these people. He had no concept
that any language could be expressed in that way. And he asked,
Where do these people come from? How can he, who sips sherry
with the great professors in Cambridge, be the product of the
same God in the same world that creates these people, so primitive? This planted a question in his mind that never went away:
How can you account for the diversity of human races?

And finally the other great experience was passing through
an earthquake in Chile. He was just sitting on the forest floor one
day and the whole earth moved beneath him. This was not only
terrifying, but it made him feel the fragility of human life. Here
he was, a young man caught in immensities he believed to be
ruled by God through natural laws. And then he reached Con-
cepción, in Chile, and he saw that the whole cathedral had been
leveled. This great house of God had been knocked down by the
same forces that elevated the Andes and changed whole geologi-
cal environments.

At the end of his life, he was asked what stuck in his mind
about his experiences in South America and on the *Beagle*. And
he remembered climbing to the peak of the Andes in Peru or
Chile—I can't remember—and then turning as he reached the

peak and looking behind him. And he said, it was like the Hallelujah Chorus in the *Messiah*, playing with full orchestra, blaring in his head, because he was on top of the world. He was looking down almost like God upon this creation, which he had begun to sort out in his own mind as he'd been climbing, as it were. At the end of his life he was asked, "What's the most extraordinary experience you had?" And he remembered climbing to the peak of the Andes. And then he slept on it, and the next day he came back to the person and he said, "No, it was the rain forest. It was sitting there and feeling that there must be more to man than the breath in his body."

From Charles Darwin's *Voyage of the Beagle*:

> Among the scenes which are deeply impressed on my mind none exceed in sublimity the primeval forests undefaced by the hand of man; whether those of Brazil, where the powers of life are predominant, or those of Tierra del Fuego, where death and decay prevail. Both are temples filled with the varied productions of the God of nature. No one can stand in these solitudes unmoved, and not feel that there is more in man than the mere breath of his body.

James Moore was raised in an American midwestern culture imprinted by the Scopes trial of 1925. That trial arose over a law in Tennessee that forbade the teaching of "any theory that denies the story of the divine creation of man as taught in the Bible and to teach instead that man has descended from a lower order of animals." The theory in question came from Charles Darwin's

The Descent of Man. This was Darwin's sequel to *The Origin of Species.* It completed his description of creation as a self-organizing progression into every plant and animal and, finally, humanity.

Moore has written that Darwin's idea of creation by evolution was a belief born of "theological humility." Darwin sensed nothing natural and benevolent in the Victorian idea of a Creator God who had fixed every condition of life once and for all at the beginning of time. And Darwin returned from the voyage of the *Beagle* to an English society that was erecting debtors' prisons and workhouses to ward off human chaos.

✱

MOORE: London was in turmoil when Darwin reached the metropolis.

TIPPETT: What year are we talking here?

MOORE: Darwin finally arrives in London in March 1837. George IV is soon to die. Queen Victoria is soon to accede to the throne. There have been crop failures. People have been flocking to the cities, trying to scrounge a living. There's terrible overcrowding. You can see this on every street corner.

Darwin's friends paid taxes to support the poor—welfare. And welfare handouts were growing year by year, as more and more people fell on hard times and flocked to the cities. What to do with the excess number of people? If you gave them food, so the theory went at that time—this was very middle-class theory—they would simply produce more babies. Pauper boys and girls could eat well enough to reproduce, and the burden on middle-class taxpayers becomes greater and greater as the years go by. The answer being given at this time was that life should be made so difficult for the recipients of welfare handouts that they don't reproduce. In other words, they go into places called workhouses. There were workhouses in the United States, in most countries. These were places where the sexes are kept separate, and any sustenance they get they have to work for. So workhouses were being built all over the country and poor people were opposing them.

There were riots in 1839. The troops were sent in later on. In 1842, Britain came closer to revolution than any other year of the nineteenth century. And the years from 1837 to 1842 were the years of Darwin's most radical thinking about humanity's place in nature. These were the years in which he kept clandestine notebooks speculating how all of the phenomena he saw around him, in society as well as in natural history, could be explained by God's laws. The central law is the law of the struggle for existence. Darwin gets this out of Whig Poor Law ideology, and Reverend Thomas Malthus in particular, an Anglican clergyman.

TIPPETT: I'd forgotten that Malthus was an Anglican clergyman.

MOORE: Yes.

TIPPETT: Malthus described how population growth would always be too great, and that it would be checked by famine and war. But he was also saying that these things were a manifestation of God's wrath.

MOORE: For Malthus, the gap between population growth and increase of food supply is God-ordained. God has ordained this tremendous fecundity amongst human beings in order to get us to till the land, to give us the incentive to feed ourselves. We're always going to have to struggle to do that. And also the incentive to restrain ourselves sexually is a law of nature, and it's for our own good. Lots of Christians believe that. Malthus believed that. And even people who weren't particularly Christian—freethinking, radical intellectuals, Darwin's friends in London—believed that, too.

Darwin seizes on this and thinks, My Lord, if it's bad for people, think how bad it is for animals and plants because they cannot exercise moral restraint, they just constantly reproduce. And he says it's a much, much worse struggle out there for everything else in the world, and what good can come of that for them? What good can come for them is progress. The struggle produces adaptations to environments. All the things that Christian preachers had talked about as glorifying God's wisdom and beneficence, Darwin said, these things are produced by a bloody, agonized, protracted struggle out there. In the end, of course, you get adaptation to environment, things swim and fly and support themselves. But scratch the surface and it's a bloody warfare.

TIPPETT: See, what's intriguing to me here is this religious idea that Darwin toppled—that everything that was had been ordained by God, fixed, not only all the forms in nature, but even the social order, including, as somebody like Malthus would come in, even the social order which was destructive, in which people died. So there's the religious talk about Darwin's legacy of how he challenged perhaps the sovereignty of God or an idea of the sovereignty of God. But he also liberated God from being responsible for inequity and suffering, in a sense. Do you know what I'm saying?

MOORE: Darwin didn't believe that God was directly responsible for each slug and snail, each catastrophe, each premature death, each—as Darwin once said—"each gnat snapped up by each swallow." God didn't ordain these things. These things were the consequences of patterns, laws, ways of going about existence, that God had established at the outset of creation, about which Darwin didn't have anything to say, really.

And in a way, you could say he gets God off the hook. On the one hand, you can admire all the tremendous adaptations and the progress in the natural world and ascribe this to laws prescribed by God. On the other hand, you have to balance out that good with the pain that we experience. Darwin doesn't offer any form of compensation. He doesn't say there's going to be a heaven for dogs or for horses or for people. He does suggest that in the future our descendants will look back upon us in the way that we look back upon the apes. They will be that much more advanced than the rest of us. And that was just a piece of Victorian optimism, you know.

TIPPETT: I wish I could say that I felt that were being proved in our time.

MOORE: There was a moment, a very poignant moment in the 1860s, when a friend of his lost a relative and wrote to him, rather distraught about the meaning of human existence and the meaning of death in this universe and how awful it is to lose a relative. And Darwin wrote back and said, hey, that's nothing compared to the death of millions of species throughout recorded history in the collapse of the solar system. And then he inserts in the letter the words "*sic transit gloria mundi* with a vengeance": and so passes the world with a vengeance. There was something deep inside Darwin that wanted to bring people face to face with the appalling depths of nature—that it produces morality, nature, but it's not a moral place. There's no comfort in nature. He grits his teeth and he makes us look at it in *The Origin of Species*. For all the God and the glorification of God's creation you find in *The Origin*, there is also this bloody-minded insistence that there are no simple solutions.

✳

A letter from Charles Darwin to Harvard botanist Asa Gray, July 3, 1860:

> I see a bird, which I want for food, take my gun and kill it. I do this designedly. An innocent and good man stands under a tree and is killed by a flash of lightning. Do you believe that God designedly killed this man? Many or most persons do believe this. I can't and don't. If you believe so, do you believe that when a swallow snaps up a gnat, that God designed that that particular swallow should snap up that particular gnat at that particular instant? I believe that the man and the gnat are in the same predicament. Yet I cannot persuade myself that electricity acts, that the tree grows, that the man aspires to loftiest conceptions, all from blind, brute force.

✳

TIPPETT: There's something that jumps out at me, and I don't see any commentary on it in anything I've read. The analogies Darwin makes, the words he uses—he drew a picture as he formulated his idea of natural selection, and it was of a tree. Here's one way he describes natural selection: "As buds give rise by growth to fresh buds, and these, if vigorous, branch out and overtop on all sides many a feebler branch, so by generation I believe it has been with the great Tree of Life, which fills with its dead and broken branches the crust of the earth, and covers the surface with its ever branching and beautiful ramifications."

Now, what intrigues me is that he uses that phrase, "the Tree of Life," which harkens back to Genesis, the tree in the center of the Garden.

MOORE: Absolutely.

TIPPETT: Was that in his mind, in his imagination?

MOORE: I have little doubt that it was in his mind.

TIPPETT: That's fascinating.

MOORE: He was not devoted to the scriptures, but he lived in a culture that was saturated with the phrases of the King James, the 1611 version of the Bible. And this tree, for Darwin, is a genealogical tree. It is the common ancestry of us all. At one point he says in his notes, "We are all netted together." Or in another note when he's a young man, "It's more humble and I believe true to see us as created from animals." That tree is the tree of how we relate to everything else that is alive. And for Darwin, that isn't to reduce human beings. It's to raise everything that grew on that tree, even the branches that fall off, the twigs that are lost. These are the things that go extinct.

TIPPETT: That wither because they go extinct, yes.

MOORE: They fall into the earth and they form the soil in which others grow. It's a wonderful vision of the richness of organic nature and the unity of life.

TIPPETT: And of human participation and belonging to that larger picture.

MOORE: Darwin has a vision of nature and it developed over a long period of time—from when he was in his twenties really until at the end of his life when he's working on earthworms, of all things. I do have the most profound respect for the way he doggedly pursued his vision of the history of life on Earth and how great things are caused by little things. Mountains move up by small increments, the soil of the Earth is recycled through earthworms, coral reefs grow by tiny increments over tens of thousands of years. No one can see these things happening. One has to be able to imagine them happening. And Darwin had that wonderful imagination. He had the capacity to sit still or stand still in a field or in a wood for an hour at a time, and just watch and listen. There are few of us who have that today, and we're the worse for it.

TIPPETT: Right. You're giving me a different way to think about one of the reactions people have in this country, that has been publicized as these things have been publicized in some court cases in recent years—evolution, intelligent design. There are people—and maybe this was a reaction some had in Darwin's time, too—who really take offense at the idea that we human beings came from monkeys. That somehow this idea diminishes what it means to be human. The way you're describing Darwin's approach is in fact exactly the opposite.

MOORE: Darwin's approach is very much in harmony with people who are against speciesism, as it's called today. Darwin abhorred cruelty to animals. He remonstrated with people who he saw abusing animals. He would take them up on it on the spot.

He was a JP, a justice of the peace, a magistrate for his county, and there are cases of him sentencing people to punishment because of the way they treated their pigs or their horses. Darwin even respected plants, and there are descriptions of him going into his greenhouse and talking to them and stroking their leaves as if they were alive. Darwin wasn't a tree hugger; I don't mean that at all. He respected life. He wasn't averse to killing animals and dissecting them. He wasn't a vegetarian, but his vision of us all being netted together— the human races as one family and all of life as part of the great tree of life whose creator, through the laws of nature, Is God—is Darwin's way of looking at the world.

TIPPETT: I wonder if you could talk about the religious reaction to and debate about Darwin's ideas in his time, and how that is similar to or different from the debate that flares up again and again in ours. Did it have the same dynamics? Did it have the same theological positions?

MOORE: No. No, it's not the same. History hasn't been repeating itself. Darwin's colleague Alfred Wallace is believed to have come up with the same theory of natural selection twenty years before Darwin did—in fact, it was Wallace's work that got Darwin to publish *The Origin of Species* quickly to establish his priority. This man Wallace, who was considerably younger, went to the United States for a lecture tour in 1886. He started off in New York, Boston, and Washington. Then he made his way across by train through Kansas, Iowa, and Nebraska, and he got to California. And during his trip, he lectured on Darwinism, but there was no problem. He was welcomed, and he got his lantern and slides out and explained Darwinism. That's what he called it, Darwinism. It shows that between 1886 and '87—when Wallace was trumpeting Darwin's cause in America—and 1926 and '27, forty years later, a remarkable change took place in the way that ordinary Americans were prepared to look at evolution.

TIPPETT: Was that the year of the Scopes trial, 1926?

MOORE: Scopes trial was 1925, but there was continued agitation even for a while after William Jennings Bryan's death.

TIPPETT: How do you explain that? What changed?

MOORE: First, a lot of people got educated, and not just about evolution. Most people didn't go to university. A lot of people got educated by their ministers, who themselves had had higher education and had come to believe that evangelical civilization was slipping away from the churches. This has to do with mass immigration from Europe, particularly of the darker skinned in Catholic parts of Europe in the 1880s and 1890s. It has something to do with the growth of cities, with labor unrest. Most important, I think the change from the 1880s to the 1920s hinges on the First World War. It was William Jennings Bryan, the great populist politician fundamentalist who went to Dayton, Tennessee, who in the manner of a political crusade brought it to the attention of Americans that German generals had quoted Darwin and Nietzsche to justify the savage campaigns of that war and the mass death.

TIPPETT: What I also hear when you describe that time, the early twentieth century, is that the details are different, but we also live in a time of tremendous change. Immigration is an issue for us. But it's not immigration anymore, it's globalization, it's transnationalism, it's a world that is changing. It's easy to be fearful and, I think, to kind of batten down the hatches.

If it is possible to make a correlation between fear of Darwin and a world that is changing—and fear of that change and things we don't understand and can't control—then I think human fear is understandable in these circumstances, and predictable. Out of everything you know about Darwin and what you've learned and even the evolution, if you will, of where you came from, your more anti-Darwin religious upbringings and where you are now, many years later in Cambridge, how would you speak to that fear?

MOORE: There's a historical philosophy underlying this form of fearful fundamentalism, which suggests a kind of conspiracy, and it's linked with Darwin and Marx and Sigmund Freud. It's linked now probably to Islamic fundamentalism, that we are fighting a malignant, invisible world. Malignant, invisible worlds are really in fashion at the moment. Think about, you know, *The Da Vinci Code*. Intelligent design, it seems to me, is the scientific equivalent of *The Da Vinci Code*. There's a mysterious intelligence behind what appears in nature, and it's very plausible that there is some evil design in this intelligence, and people believe it.

TIPPETT: And that that's been covered up also.

MOORE: And it's been covered up, that's right. So, you know, as I was brought up, some of the intellectual influences in my life were of that conspiratorial nature, that really the Earth is a sinking ship, that there's nothing much we can do about it except to get people into the lifeboats. I don't think that's the dominant impulse today in Western fundamental or evangelical Christendom. It's much more of a conquering and triumphal spirit. But also one that must struggle with God's enemies.

TIPPETT: But I also think you do not conclude that Darwin was an enemy of God. That's not a place you've come out.

MOORE: Absolutely not. I didn't know for sure whether Darwin was an enemy of God when I started out. I was given to believe that he was, at best, a well-meaning man, at worst, a sort of demonic figure. It became clear to me that he was not a professional theologian or a philosopher, for sure. But he was a very shrewd guy, and he'd stared more deeply into the abyss, which is his view of nature at war, than perhaps any person of his day. And he brings you up short, bang, against the world as it really is in his vision, not the world that we would like it to be, as if there hadn't been a fall into sin in the Garden of Eden.

5

*

Content with the Limits of
Religion and Science

"THE HEART'S REASON"

This conversation with V. V. Raman provides an intriguing glimpse into the rich global dialogue between science and religion that is obscured by Western headlines about a science-religion clash. It also serves as a kind of introduction to Hinduism, a tradition that shapes one billion people, most of them in India. Deeply rooted in Indian culture, Hinduism was identified only in the nineteenth century, by European scholars, as a formal religion. V. V. Raman prefers to describe it as "a cultural religious worldview that has given rise to an impressive body of sacred literature, magnificent art, great music, majestic architecture, and profound philosophy."

As vitally as any other tradition, Hinduism has kept an awareness and practice of art as life-giving at the very center of daily lived spirituality. V. V. Raman's words in this conversation and the readings that accompany them convey some sense of this.

This overarching regard for beauty is not unrelated to the fact that Hinduism has historically avoided a point-counterpoint between science and religion. It is a reflection of a core Hindu insight that multiple forms of knowledge have a place in human life. In V. V. Raman's mother tongue of Tamil, language itself distinguishes between the word "why" as a causative question—the way science approaches a problem——and "why" as an investigation of purpose——the way religion might approach the same problem, with very different results.

As V. V. Raman sees it, knowledge conveyed by art and poetry and beauty is not "irrational" but is "transrational"—and as critical in human life as rationality. He uses the analogy of a sonnet. Logic can analyze it powerfully in terms of structure; the human spirit will plumb it for meaning. He juxtaposes shared elements of both science and religion to explore the complementarity of these two realms of human endeavor. He's written intriguingly, for example, about "numbers" in science and religion. He experiences the multitude of deities in Hindu spirituality as an expression of the kindred insight of science and religion that there are no simple answers to complex questions.

Ultimately, V. V. Raman is also content with the limits of both science and religion, and the room they leave separately and together for mystery. Karma and reincarnation, for example, are not concepts he would defend with his scientific colleagues, but neither does he believe that they can claim any more authoritative convictions about "postmortem existence." Where our culture assumes cognitive dissonance, Raman says, he consistently arrives at an "experiential consonance." I suspect that this consonance is experienced by many of us.

*

The Heart's Reason

KRISTA TIPPETT, host
V. V. RAMAN, theoretical physicist and author

$\mathcal{V}$ aradaraja V. Raman is emeritus professor of physics and humanities at the Rochester Institute of Technology in New York. He has been described as "a transcultural voyager . . . who . . . courses from physics to philosophy, from music to metaphysics." He's lived and taught in the United States for four decades, but he was born into a Brahmin Tamil family in Calcutta in 1932, and educated in mathematics and physics in India and Paris. Raman has devoted his life to the science of physics and to the elucidation of Hindu religion. He has come to regard himself, he has noted, "as an inheritor of two great traditions, as I see it: one, the Hindu tradition on the religious plane, and the other, the scientific tradition, which I regard as one of the greatest intellectual and spiritual triumphs in the history of humankind."

V. V. Raman has helped to edit an eighteen-volume encyclopedia of Hinduism. He's authored scores of papers on the historical, social, and philosophical aspects of physics, as well as on the heritage of his native India. His books include *Glimpses of Ancient Science and Scientists*, and it is from his long imagination about history and time that V. V. Raman begins to put into perspective a Western sense of science and religion at odds. Modern science emerged in western Europe, he says, and its immediate discoveries contradicted specific church teachings. But this kind of point-counterpoint never happened in the Hindu world.

RAMAN: In the Hindu world, fortunately, there was a clear understanding of what constitutes religious knowledge on the one hand, and what may be called intellectual, analytical, secular knowledge. This distinction is much more clear, it seems to me, in the Hindu world, which is why we don't have this kind of conflict.

TIPPETT: So in your way of seeing the world, then, as a Hindu, is there never a conflict? There's a distinction and yet not a divide?

RAMAN: Exactly. One often talks about cognitive dissonance, for example. Now, I rather call it an experiential consonance. And what I mean by that is that it is possible to distinguish between what we understand and explain in the logical and analytical framework, which is what science provides. And to distinguish that from another level of experience in the world, which comes from what may be called deep involvement. It is not unlike enjoying music on the one hand and then proving a geometrical theorem. You can do both.

These are two kinds of experience, and the human spirit, if I may use the word, and the human dimension is so complex, that we have all kinds of possibilities. One of the unfortunate consequences of the successes of the sciences is this addiction, as it were, to rationality.

TIPPETT: An addiction to rationality.

RAMAN: By which I mean that every single aspect of human experience must be subjected to rigid rationality. Now, I have the greatest respect for reason and rationality. But I also think of the Ecclesiastics, who may say, "To everything there is a season and a time to every purpose under heaven," which has been articulated by thinkers through the ages in all the cultures, I would say. When Pascal wrote his famous statement *"Le cœur a ses raisons que la raison ne connaît point"*—the heart has its reasons which reason doesn't understand—those are ways by which the enlightened thinkers and visionaries understood that the world is far too complex for us to really rigidly put everything under the strait-jacket of reason.

TIPPETT: You make a point in something you've written that reflects an observation I've made. So much of our cultural debate about science and religion seems to assume that science and religion pose competing answers to the same questions. But in fact they pose different questions. And you note that in Tamil there's a distinction linguistically between "why" as a causative question, the way science might ask why of a problem, and "why" as a teleological question the way religion might ask it. I thought that was very interesting.

RAMAN: I think it's a very, very important distinction. Both kinds of "why" are important, in that the human mind cannot escape those questions.

TIPPETT: And we start asking those questions from a very young age, don't we?

RAMAN: A very young age. But the languages influence our way of thinking. I sometimes ask my students, "Why are you taking this course?" Some students may say, "Because it is required in my curriculum." Others may say, "Because I want to learn what you are going to talk about." Now these two answers are legitimate answers to the same question. The first answer implies a framework in which the student is operating. But the second is purposeful and teleological: "Because I want to learn." It's in the future, whereas the first one is because that's how the rules are set up. So both questions are relevant and interesting. Except that as I see it, the question about "why" in the deeper sense of what is the purpose of this universe—Why am I here? Why was the world created at all? Why are the laws such as they are?—those are very fundamental questions for which we may never be able to find answers which are unanimously acceptable.

✳

Hinduism is the world's third-largest religion, after Christianity and Islam, but it is by far the most ancient, as is its sacred language of Sanskrit. Alone among the world's major traditions, though, Hinduism has no known founder and no identifiable point of origin in history.

V. V. Raman has called Hinduism "a cultural religious worldview that has given rise to an impressive body of sacred literature, magnificent art, great music, majestic architecture, and profound philosophy." Its foundational truths are captured in ancient scriptures, known as Vedas, and conveyed by epic poetry and saga such as the Bhagavad-Gita. One of V. V. Raman's starting points for imagining the compatibility of science and religion is the impulse of universality that he finds both at the heart of science and in Hindu spirituality.

✳

TIPPETT: I think it is striking that although Hinduism is the third-largest world religion after Christianity and Islam, it's the least known. It's the least in the headlines, partly for positive reasons, because it's not making the news in negative ways these days.

RAMAN: Yes.

TIPPETT: But it's not as well known in U.S. culture even as Buddhism, which grew of Hinduism. If people have an image at all in their heads, it is of this multitude of deities. And that does not evoke universality, nor does it evoke a religion that is compatible with logical thinking. So talk to me about how you respond to those kinds of stereotyped images that are out there. Or partial images, let's say.

RAMAN: Sure. I think there is every reason for that misunderstanding. One of the fundamental scriptures of Hinduism is known as the Vedas, the Rig Veda for instance. And in the Rig Veda, the most important aphorism or statement is "Truth is one and the people call it by different names." In Sanskrit, the word "truth" is *sat*—it's called *ekam sat*: "there is but one truth." I like to look at it as follows: How many music are there? Even the question doesn't sound right. In order for anybody to understand or appreciate music, one can only do it in terms of a particular song or sonata or concert or . . .

TIPPETT: Or a genre . . .

RAMAN: Or any genre of music, and a particular piece specifically. Now, the Hindu gods are, to me, somewhat like different pieces of music.

TIPPETT: Of a variety of melody and tempo.

RAMAN: . . . the sheer variety. Probably everybody has their own favorite music, favorite piece. Likewise, in the Hindu world there is something called a favorite god, believe it or not. It's called *ishta devata*.

TIPPETT: People tend to identify very strongly with a particular god.

RAMAN: Yes, they have a special regard for that particular depiction of the intangible. Every god is simply a representation. They are not any different, if you want to give an analogy, than having different saints in the Catholic tradition who are worshipped on different days, for example.

TIPPETT: You also write about a fundamental insight of Hinduism that also finds expression in this multiplicity of tradition and gods, this fundamental insight that there are no simple answers to complex questions. That's an important insight for our time, in every sphere of life.

RAMAN: In fact, my own personal view is that religious experience is precisely in the experience of that mystery. There is in human life a certain mystery surrounding all this. And it is the experience of that mystery—even if it is only momentary and even if it is only for a few minutes every day as, for example, when I do my meditation or whatever—that is what constitutes the religious experience. As soon as we unravel that mystery in words and in formulations, it becomes the doctrine of a religion. Many of the religious doctrines are profound answers to the mysteries, but they become interesting and important more in historical and geographical terms than in ultimate terms.

TIPPETT: You're saying that that experience of mystery always in some sense eludes and transcends the doctrine that it became?

RAMAN: Absolutely. And a doctrine may answer, within a religious framework, some of the mysteries. To the extent that it gives fulfillment to the practitioners, I have no problems with that. Even taking that to be universal, again, is not wrong as long as one does not impose that on other people who may have different answers to the mysteries.

*

Here's a passage from V. V. Raman's 1997 book about the Bhagavad-Gita:

> The most important realization of Hindu seers, the fundamental revelation that comes from their meditation and spiritual search, is that beneath and beyond the material and the physical world lies a spiritual reality. It's only when one recognizes this that one has truly lived the human life.
>
> An analogy with the physicist's endeavor may clarify this thesis. We see, observe, and experience countless physical phenomena around us: lightning and sunrise, erosion of rocks and the colors of the rainbow, the blossoming of flowers and the freezing of water in the cold, and many more. But when we become aware of these as various consequences of fundamental physical laws, our depth of understanding is enhanced and our appreciation of the phenomenal world is enormously enriched. Likewise, say the seers, when we become aware of the spiritual substratum of the universe, our experience of it is heightened a thousandfold. Indeed, it is only when we achieve this that we really begin to see—that is, to understand—anything.

✳

TIPPETT: I'd like to ask you about some key ideas in Hinduism and what they mean to you, also how you live them and experience them as a scientist. And one of those is *karma*. I'd love to know what karma means for you. How do you reconcile that kind of idea with what you know as a physicist?

RAMAN: Okay. An associated word which I think is equally important in the Hindu world and which has come into the West with different connotations, is "dharma."

TIPPETT: Dharma, yes.

RAMAN: Very simplistically put, dharma is what we are expected to do and karma is what we do. Dharma has been translated variously as "duty," as "religion," as an ethical framework. And there are many treatises in classical Hinduism which talk of dharma in different ways. One of them, for example, lists such things as mercy and temperance, adherence to logic, the pursuit of knowledge, the pursuit of truth, not getting angry. These are some of the kinds of ethical principles . . .

TIPPETT: An "essential virtue" is what comes to mind . . .

RAMAN: The essential virtues. The dharma which is set to be the crucial one is the pursuit of truth. And that means everything from being kind to others, being respectful to parents, those kinds of things. Now, karma is a metaphysical concept and the Hindu answer to what is sometimes called the problem of evil.

TIPPETT: Karma is a response to the problem of evil.

RAMAN: Yes. Evil and the challenge it poses to religious faith—theodicy—the question, how can you say that God is just and good and kind in the face of earthquakes and natural disasters? Different cultures have come up with different answers. The Hindu answer is that evil in the sense of suffering is ultimately a consequence of one's own actions. So karma is any consequential action, any action that has an impact, positive or negative, on yourself or on others.

TIPPETT: And implicit in that is a belief in reincarnation or in many lives, that life is not this linear, one-time thing.

RAMAN: Absolutely. We cannot explain that. We talk of people getting away with murder. The Hindu idea is—not forever.

TIPPETT: Though you might get away with murder in the moment.

RAMAN: This time. But the idea of transmigration, reincarnation, is inevitable in the framework of karma. Now, the way I interpret the karma doctrine is as follows: at the very least, it makes one take responsibility for one's suffering, rather than point a finger at someone else.

TIPPETT: And is the idea that though you are living with the consequences of previous actions, the way you live this life could determine a better future?

RAMAN: Absolutely.

TIPPETT: I want to know, though, how you think about that, how you hold that belief, with everything you know about physics and cosmology as a scientist. Would you be able to talk about that with a fellow scientist in a way that would seem legitimate to them?

RAMAN: No. I don't think I can argue for reincarnation from a scientific perspective, quite honestly. There are people who have done research on this question and who have quoted cases where people have vague memories of past lives and all that. I have to confess that as a physicist I will leave that open. I do not have any firm convictions as to the mysteries of postmortem existence. See, I take that as one of the mysteries for which I don't know the answer. And I rather suspect others who claim to know. But modifying Hamlet slightly, I would say there are more things in heaven and Earth than are dreamt of in our sciences.

✳

For many years, V. V. Raman has written frequent short essays for friends and colleagues on art, religion, and science. He's reflected on diverse religious figures, world leaders, and history ancient and modern. Here are some lines from one personal essay Raman sent to friends and colleagues:

> We use words to talk. We enjoy music. We play with numbers. In the Hindu framework, there is a goddess who gives us words and language and music and numbers. That goddess is called Sarasvati. Today the Hindu world celebrates that name joyously and ceremoniously. By tradition, we are not allowed to read today. Books in the house are placed on a pedestal and worshipped. But tomorrow, at crack of dawn, children are expected to rise early from bed and read from a book, with a resolve to do that every day of the year.

✳

TIPPETT: You write about how in the Hindu framework there's a goddess who gives words and language and music and numbers, Sarasvati?

RAMAN: Yes.

TIPPETT: So talk to me about how you live with a piece of mythology like that and live with what you know again about the physical universe, about numbers especially.

RAMAN: See, mythology has become a fairy-tale sort of word.

TIPPETT: Right. The implication in our culture is it's something that's not true. I don't use the word that way, but yes.

RAMAN: I would be the first to say that this is part of Hindu mythology. But there is something called "mythopoesis." These are parts of all the great religions of the world. The poetic aspect is extremely important to me, because poetry is what gives meaning to existence. Not fact and figures and charts, but poetry. Poetry is essentially a really sophisticated way of experiencing the world. And it is much more than mere words and stories. Poetry is to the human condition what the telescope and the microscope are to the scientist.

So I do a meditation to Sarasvati. There are images of Sarasvati, very beautiful, beautifully clothed in a sari and with a vena, the grand musical instrument of India, and a rosary, which corresponds to the counting, the numbers. To me, this is imagery that evokes reverence and respect, not necessarily for the particular form in which it is depicted, but for all those intangibles, such as counting and numbers and music and knowledge and science, which enrich human life and human culture and human civilization. It's an aesthetic experience to contemplate on something symbolic like that. I'm well aware that ultimately all these are symbols and that they may not reflect exactly what is out there. But we live in symbols as long as we are cultural beings, and that is how I take it.

I remember we used to do a prayer to Sarasvati in school every morning. Even now I think there are many schools in India which do that. And somehow it inspired us to go through the days of learning. It hasn't, quite frankly, done me any harm. What I mean by that is I'm amazed at the kind of objections people raise to having a moment of prayer in school in America now. Believe it or not, I also went for some time to a Jesuit school and I repeated Paternoster in Latin; that didn't do me any harm either. As far as I can see, these are inspired. These are parts of great traditions, and they can only infuse reverence and respect in the hearts of children.

TIPPETT: You've noted that there's a fascinating importance of numbers in both science and religion. I'd like you to say something about that. It is quite interesting when you start thinking about it.

RAMAN: Numbers, as you know, are in some ways mischievous. Although we concretize them when we count objects and things and days and hours and so on, we talk of numbers always in reference to those. This is another example of polytheism, if you like, because nobody can image numbers except in those concrete terms of counting. Numbers themselves are far more abstract, and philosophers of mathematics have often wondered and argued about whether numbers like the so-called irrational numbers and transcendental numbers and transfinite numbers have any reality.

TIPPETT: Numbers becomes quite mysterious, don't they?

RAMAN: They become mysterious. And my own feeling is that may have been a reason why, one way or another, the religious traditions of humankind have incorporated numbers in specific ways. In the scientific world, numbers play a very, very different role. They are again associated more with natural phenomena.

TIPPETT: I've always been fascinated in my conversations with scientists about how they find great beauty in mathematics.

RAMAN: Absolutely. Though that is more, much more than numbers.

TIPPETT: It's almost rapture. Right, it's more than numbers.

RAMAN: But you are absolutely right. For the mathematician or for the physicist, the idea of mathematics . . . I think it was Sir James Jeans who said that God, for want of a better word, may be called "mathematical thought" or something like that. Because ultimately it is the mathematical beauty of the universe that grabs the physicist especially. Maybe not all scientists, but physicists. There is something aesthetic about the laws of electromagnetism, for example, formulated by Maxwell or the so-called direct equations, and on and on. That is very true.

Here's an excerpt of V. V. Raman's essay "Numbers in Religion."

Every major religion refers to numbers and attaches particular significance to certain numbers. In Egyptian religion, there was "number mysticism." The number 3 takes on a special significance in many religious contexts: Anu, Bel, Ea in mesopotamia; Isis, Osiris, Horus in ancient Greece; Brahma, Vishnu, Siva in the Hindu tradition; Father, Son, and Holy Spirit in Christianity, and so on. Four was important in ancient recognitions of elements: earth, water, air, and fire. In Chinese lore, 5 is the important number. The Pythagoreans regarded 6 as the perfect number because its factors, 1, 2, and 3, also add up to it. In the Judaic tradition, numbers are associated with Hebrew letters, and this enables experts to uncover esoteric meanings in words. The ancient Babylonians recognized seven celestial bodies that moved differently than all the stars in the heaven. Islamic scholars point out that the Qur'an's magic number is 19. Buddhism speaks of the 12 golden rules, Jacob and Ishmael had 12 sons, Elijah built an altar of 12 stones, and Christ has 12 apostles, etc. Thus, numbers come into religious contexts in many instances. Could this be because numbers are as abstract as God and as relevant to human life as religion?

TIPPETT: We've been talking about how your religious sensibility relates to your scientific sensibility. You talked about how karma is a Hindu response to the problem of evil. I wonder if your scientific knowledge and perspective also informs something like, let's say, the way you think about the problem of evil in human life and even evil within religious traditions. Are there things you know as a physicist that give you more to work with, as you make sense of that personally?

RAMAN: Certainly, I think my involvement in physics and the sciences has given me an historical cultural understanding of many of these enormously meaningful things in life. Because science, among other things, enables us to look at human events in human terms. Religions, in their context, enable us to look at human events in religious or transrational terms. Both, in a way, are meaningful and illuminating. When you read a sonnet, let us say. Science is the discovery of the rules of prosody, the rules by which the sonnet is constructed, of measure and syllable and accent, iambic pentameter or whatever.

TIPPETT: Right.

RAMAN: You can analyze a poem, and this understanding of the structure of the poem is a significant accomplishment. But it tells us nothing about the meaning behind the poem or about the inspiration that the poem might give. And the universe, to me, is somewhat like that. Science enables us to understand the laws and principles by which the universe is constructed, its functions. That is no trivial accomplishment. One of the greatest intellectual achievements of the human mind is what modern science has been able to do. But there is always the question of meaning. And while it is possible to derive meaning without going beyond the physical world—and many people do it—it is no less inspiring and fulfilling to find meaning within a religious framework insofar as it is not irrational. There's a difference between irrationality and transrationality. To me, many of the deeper messages of religions, such as caring and compassion and respect for others, helping others, love, reverence, these are not rational. They are not irrational. They are transrational and they have their sources in the many religious frameworks of humankind. They not only carry the weight of centuries, they also reflect something deep in the human cultural psyche.

TIPPETT: And yet, as you know, we unfortunately don't always just see the best of religions. Language about what is "trans-rational" carries a new sense of threat in our time. There is a great deal of violence being committed in the name of God and transcendence. How do you watch that and how do you think about that?

RAMAN: That is a perennial problem. I have tried, naturally, to articulate whatever is the best and illuminating in the religious traditions, if only because there is ample evidence of whatever is worst in the daily news.

TIPPETT: Nobody needs to articulate that, right.

RAMAN: And it is depressing that we live in an age when religions have become associated with politics and violence and war and recriminations. If anybody is to grieve for this, it should be the gods above, because this is not what religions were meant to be. And it is true that in this context it is extremely important for the leaders, the intellectuals and the thinkers of the world, to speak out openly about all that is bad and evil that has come out of religions. But given that religion is such an intrinsic part of human culture and means so much to at least four billion, perhaps five billion human beings, we must ask this: what can realistically be done, if one may use the term, to tame or bring out whatever is still good and worthy in religions?

TIPPETT: Perhaps one can say that a dark side of Hinduism, and which seems to defy the virtue of universality, is the caste system.

RAMAN: That is a very important point. I don't want to be apologetic here; I will be the first to say—and I am part of a growing number of Hindus, both in India and abroad, who are speaking out and writing against the evils of the caste system.

TIPPETT: Okay.

RAMAN: But the point to remember is that casteism is a slowly but surely disappearing aspect of Hinduism. And all through India's history there have been so many poets and thinkers and philosophers who have spoken out against what can only be called the scourge of casteism.

It is a slow ingrained process. Personally as a Hindu, I will say that I have never been pleased with casteism being part of my own religion. And although I was born in a Brahmin family, I refused to accept the caste title that goes with my name. It is not something that can be defended in any way in the modern world. The world has changed in many ways, and so does Hinduism, as it ought to, as all civilized religions ought to.

TIPPETT: In the last century there is a person who almost embodied Hinduism for many, and that would have been Gandhi. Gandhi is still this amazing figure who influenced leaders of other religious traditions and was even revered by Einstein.

RAMAN: I belong to a generation which worshipped Gandhi. In high school I attended mass meetings where Gandhi spoke. Several hundred thousand people were there. Gandhi was extraordinary in many, many ways. Most of all, he understood that basically human beings are decent. And that no matter what, it is by trying to bring out whatever is good and noble in the human personality that we can resolve many complex problems.

Now, I will be the first to grant that this can be idealistic talk and in many instances it simply may not work. People have pointed out, could you have applied nonviolence to Hitler and so on. But we need to strive for or at least try to see if we can resolve problems by peaceful means and by trying to be understanding of the opponent's point of view. That is the key. Gandhi is a supreme example, and I'm glad there were people like Martin Luther King and Nelson Mandela, two outstanding people in later times who followed Gandhi's path. There is really little hope that we can resolve the complex problems of the world by continuing to escalate anger and hatred, however justified it may seem from one's own perspective.

TIPPETT: And that's, for you, the important legacy of Gandhi right now.

RAMAN: I think so. And I think Gandhi has become extraordinarily relevant. I said I belong to a certain generation. Gandhi is not so highly regarded today in many parts even of India because of all the frustrations and chaos caused partly by his excessive effort to understand the opponent. There are people who have argued that it is that attitude which has resulted in . . .

TIPPETT: Has created problems.

RAMAN: Yes. We don't know. But I think we can never give up ideals if civilization is to last.

❋

Here in closing is a passage from V. V. Raman's writings about the great Indian poet Rabindranath Tagore. Tagore influenced Mahatma Gandhi, and he won the Nobel Prize for Literature in 1913.

Tagore was a prolific writer, musical composer, artist, but, above all, a Bengali poet par excellence. He was gifted, through some mysterious genetic coding, with rhyme and rhythm, with inner melody and exuberant creativity. In his offerings, Tagore reflected on the inner essence of reality and there first appeared his immortal lines, "Where the mind is without fear and the head is held high; where knowledge is free; where the world has not been broken up into fragments by narrow domestic walls; into that heaven wake this Indian land." If Tagore was profoundly moved by the glorious insights of Upanishadic texts, he was no less appalled and pained by the inhumanity of casteism and the mindless mutterings of heartless orthodoxy. The perennial prayer of ancient India, the vibrant theme that is echoed all through Indian history, is also given due place, for the poet pleads: "Oh, grant me the prayer that I may never lose the bliss of the touch of the one in the play of the many." It is in the words of the poets that the deepest religious feelings of humankind survive.

6

※

The World Feels More Spacious

"MATHEMATICS, PURPOSE, AND TRUTH"

I picked up Janna Levin's novel *A Madman Dreams of Turing Machines* off a table at a bookstore. I was drawn to it initially because we had just completed a program on autism, in which Alan Turing —known as the father of modern computing—was one historical figure discussed. I was immediately taken by Levin's lush prose and the alluring, provocative ideas that she brings to life through human stories in space and time.

A Madman Dreams of Turing Machines sounds depths I had never considered before, delving into mathematical truths and great existential questions. It does so by probing the parallel lives and ideas of Turing and another pivotal twentieth-century mathematician, Kurt Gödel. Turing's discoveries were made possible in part by Gödel, who shook the worlds of mathematics, philosophy, and logic in 1931 with his "incompleteness theorems." They demonstrated that some mathematical truths can never be proven. Or, as Gödel says in Levin's novel, "Mathematics is perfect. But it is not complete. To see some truths you must stand outside and look in." This held unsettling scientific and human implications; it posited hard limits to what we can ever logically, definitively know.

Janna Levin is an atheist, if we care to categorize her. And while that simple fact informs our conversation along with her exquisite intelligence and her mathematical training, we cover territory that can't be bounded by such definitions. Levin's most certain "faith" is in the conviction that we can agree on basic realities described by mathematics—that one plus one will always equal two. Putting God into that equation, or barring God from it, is not her concern. Yet this conversation is a beautiful example of the deep complementarity of religious and scientific questions, if not of answers. The ideas and questions Janna Levin lives and breathes open my mind to new ways of wondering about purpose, meaning, and ultimate reality.

There is much in her thought that I struggle to comprehend and will continue to ponder. I'm intrigued, at the same time, by echoes of the wisdom of ordinary life. Gödel's idea that there are some truths we can see only at an angle—by standing outside, looking in—is a fact even in the work I do. The deepest truths are usually impossible to see and articulate straight on.

And I feel a kindred pull to Levin's delight and passion in the great narrative of the world and humanity, epitomized in these lines from her book:

> I am looking on benches and streets, in logic and code. I am looking in the form of truth stripped to the bone. Truth that lives independently of us, that exists out there in the world. Hard and unsentimental. I am ready to accept truth no matter how alarming it turns out to be. Even if it proves incompleteness and the limits of human reason. Even if it proves we are not free.

Of all the ideas Levin presents, the most provocative and disturbing, perhaps, is her doubt that there is free will in human existence at all. She cannot be sure that we are not utterly determined by brilliant principles of physics and biology. Yet she cleaves more fiercely in the face of this belief to the reality of her love of her children and her hopes and dreams for them. She sees "evidence of our purpose" in figures like Gödel and Turing, even though they did not the find the clarity in life that they wrested from mathematics on all our behalf.

Paradoxically, perhaps, the world feels more spacious to me after this conversation with Janna Levin—even, to use her words, if it suggests incompleteness and the limits of human reason and faith; even if it suggests we are not free. She possesses a quality that keeps me interviewing scientists as often as I can —a delight in beauty, a comfort with mystery, a limitless ambition for one's grandest ideas combined with a humility about them that many religious people could learn from.

Mathematics, Purpose, and Truth

KRISTA TIPPETT, host
JANNA LEVIN, physicist and novelist

✳

J anna Levin is a theoretical physicist with a special interest in the origins and shape of the universe. She is a professor of physics and astronomy at Barnard College. She's also the author of a novel, *A Madman Dreams of Turing Machines*, that explores great existential questions by probing the lives and ideas of two pivotal twentieth-century mathematicians, Kurt Gödel and Alan Turing. Turing is known as the father of modern computing, and his insights were made possible in part by Gödel's discoveries. Janna Levin's novel imaginatively evokes the force of their ideal in the classrooms and coffeehouses of Gödel and Turing's day, and in her own life as a twenty-first-century urban scientist—though she tells me she began her undergraduate studies with little active interest in science, convinced instead that philosophy was asking all the big questions.

LEVIN: It's very ironic, when I look back at my childhood, that I was absolutely mesmerized by cosmology and astronomy, even evolutionary science, ideas on natural selection. They had always captured my imagination with these gratifying ways to think about the world. Even if I didn't always understand the answers, it was a way to think about the world.

TIPPETT: Tell me how you made that transition when you went to college and you were studying philosophy. How did you get captured by theoretical physics?

LEVIN: I hadn't really admitted to myself that I love science. And then I was in a philosophy class, and I was impressed with the subject. We were talking about a lot of interesting things—free will, indeterminism, what it means to say we're free in a world that's completely, causally, physically determined. These are very deep questions. And one day, a scientist came in to give a guest lecture and started to discuss quantum mechanics. Everybody in the room got very quiet. They discussed Einstein. And what I was most impressed with is that philosophers didn't know how to respond. I thought this was powerful, and I became interested in physics.

TIPPETT: This book you've written about Kurt Gödel and Alan Turing takes place very much at that intersection where philosophical questions meet scientific inquiry and scientific truth.

LEVIN: Yeah, I definitely came back round again.

TIPPETT: Did you?

LEVIN: In some sense I came full circle again, to start asking those philosophical questions.

TIPPETT: This basic question—let's start with Kurt Gödel, about truth, right? I want you to put this into your own words because I can't say that I can completely wrap my mind around it, but I'm utterly intrigued with it: that truth would ultimately elude us. That some mathematical truths can't be proven within the realm of mathematics—which doesn't necessarily mean they're not true, but mathematics itself can't demonstrate their truth.

LEVIN: That's right. It was a time in history when most mathematicians, I think it would be fair to say, believed that mathematics could address every mathematical proposition. That's a fair enough thing to believe in retrospect. Why shouldn't mathematics be able to prove every true mathematical fact? So when Gödel came along and found a very surreal kind of tangle, a mathematical proposition that makes a peculiar claim about itself, which cannot be proven within the context of arithmetic—it was in the context of arithmetic that he did this—it really shocked people. It really shook them up.

And I think the way he said it is actually the clearest and nicest way to say it: "There are some truths that can never be proven to be true." It opens up this idea—which terrified people—that there are limits to what we can ever know. And it's not the first time this happened. If you think about Einstein's theory of special relativity, it was a similar idea. There are limits to how fast we can ever travel. We are limited by the speed of light. There are limits in quantum mechanics to how much we can ever really know. There are fundamental limits to certainty. And we accepted all this around the same period.

TIPPETT: You have scenes with Gödel in Vienna, early 1930s Vienna, in a coffeehouse, in a famous intellectual gathering called the Vienna Circle. There's a scene where you have this mathematician, Olga Hahn-Neurath, and her husband, Otto, who's a socialist—these are just some of the people. Moritz Schlick was a philosopher and a logician who kind of headed this. And they often come back to Wittgenstein's premise—his first premise in his famous *Tractatus*—that "the world is all that is the case," which is a statement about a basic thing that we can know as real. And you have a moment where Gödel challenges this. He has been thinking about this and coming up with this theory that you describe narratively like this:

On every previous Thursday, Kurt has been a silent spectator. Tonight he looks from one person to another as he waits for the right opportunity. His temperature fluctuates while openings come and go until he throws out a question he knows they have asked themselves a thousand times. "How do you recognize a fact of the world?"

Moritz laughs, but not rudely, and nods, which loosens his hair only marginally from its proper place before he stops himself, slightly sorry for his reaction as he takes in Kurt's serious expression. "It is a fair question," he confesses. "How do I verify a fact of the world?" Such a simple question . . .

Being honest he can be sure only he *sees*. He can be sure only he *touches*. He watches Olga pull on a mammoth cigar. She has a calm about her, always at ease. The smoke drifts in curly plumes sifting through her lashes. She doesn't seem to mind and even tends to hold the burning cinder vertically and uncomfortably close to her eyes . . .

But what really arrests Moritz, what keeps his fingers in a frozen clutch around the cup of coffee suspended near his chin, is this question: Does *Olga* exist? He hangs there for what seems like a very long while. The conversation stalls, suspended along with the coffee.

"Olga?"

"Yes, Moritz, I'm here."

She reaches over and hooks his thumb with her fore-finger. The rest of her fingers scramble over to clasp his hand. But all Moritz concedes is that he can feel what he has learned to describe as pressure on what he believes to be his hand.

TIPPETT: In the novel, all the members of the circle who were sitting at the table start to question almost whether they themselves are real, whether the person who's sitting across the table from them is real. And as a reader, I had that same experience.

LEVIN: That's beautiful.

TIPPETT: It's wonderful. And so I wonder if you would describe that scene the way you envisioned it. What's happening there for you?

LEVIN: Well, I really hoped that the reader would have that experience, because ultimately I think that's where the book nudges: do you know that any of this is real, that the book isn't a figment of your imagination somehow?

TIPPETT: Even the book itself?

LEVIN: The book itself. That somehow you aren't the author of the book itself. I was definitely pushing on that limit of what do we know and what don't we know, what do we take to be faith, what's rational to believe, what's not rational to believe? And I realized that what I was writing about wasn't so much about mathematics. What I was really writing about, which I think you've struck on, is belief—what Gödel believed, what the people in the Vienna Circle believed, how they all ultimately struggled with different ideas about reality. And that there is a surreal vagueness to our conclusions.

TIPPETT: You write of both Gödel and Turing that they were besotted with mathematics. And I have to say that I feel that you—I don't know if you're completely like them in that way, but you have a real sympathy for that. You seem to delight in the way they live with mathematics and wrestle with it. Is that true? For you, are numbers maybe not "more real than the sun and the earth," but as real as the Sun and the Earth? And if so what does that mean exactly? How would you explain that?

LEVIN: I would absolutely say I am also besotted with mathematics. I don't worry about what's real and not real in the way that maybe Gödel did. I think what Turing did, which was so beautiful, was to have a very practical approach. He believed that life was, in a way, simple. You could relate to mathematics in a concrete and practical way. It wasn't about surreal, abstract theories. And that's why Turing is the one who invents the computer, because he thinks so practically. He can imagine a machine that adds and subtracts, a machine that performs the mathematical operations that the mind performs. The modern computers that we have now are these very practical machines that are built on those ideas. So I would say that like Turing, I am absolutely struck with the power of mathematics, and that's why I'm a theoretical physicist. If I want to answer questions, I love that we can all share the mathematical answers. It's not about me trying to convince you of what I believe or of my perspective or of my assumptions. We can all agree that one plus one is two, and we can all make calculations that come out to be the same, whether you're from India or Pakistan or Oklahoma, we all have that in common. There's something about that that's deeply moving to me and that makes mathematics pure and special. And yet I'm able to have a more practical attitude about it, which is that, well, we can build machines this way. There is a physical reality that we can relate to using mathematics.

TIPPETT: I want to pose a question to you that you pose in different ways to Turing and Gödel, or you have them contemplate in the novel. I'll say it this way: in your mind, does the fact that one plus one equals two have anything to do with God?

LEVIN: Are you asking me that question?

TIPPETT: I'm asking you that question. I'm asking you how you think about that.

LEVIN: I think it's . . . I am . . . oh, you're tough. I think that it raises—if I were to ever lean towards spiritual thinking or religious thinking, it would be in that way. It would be, why is it that there is this abstract mathematics that guides the universe? The universe is remarkable because we can understand it. That's what's remarkable. All the other things are remarkable, too. It's really, really astounding that these little creatures on this little planet that seem totally insignificant in the middle of nowhere can look back over the fourteen-billion-year history of the universe and understand so much and in such a short time.

So that is where I would get a sense, again, of meaning and of purpose and of beauty and of being integrated with the universe so that it doesn't feel hopeless and meaningless. Now, I don't personally invoke a God to do that, but I can't say that mathematics would disprove the existence of God either. It's just one of those things where over and over again, you come to that point where some people will make that leap and say, "I believe that God initiated this and then stepped away, and the rest was this beautiful mathematical unfolding." And others will say, "Well, as far back as it goes, there seem to be these mathematical structures. And I don't feel the need to conjure up any other entity." And I fall into that camp, and without feeling despair or dissatisfaction.

❋

Einstein described humanity's ordinary, daily sense of time as a linear, progressive arrow, a stubbornly persistent illusion. And Janna Levin's novel, *A Madman Dreams of Turing Machines*, is structured to evoke time the way physicists know it—as relative and curved, with past, present, and future in a fluid interplay. Levin occasionally brings herself, the narrator, into her fictional retelling of past events, commenting on them from modern-day New York City. Here's one such passage:

> I have tried to stay out of these stories but I am out here too. I am standing on a street in a city . . .
>
> In the park, over the low wall, there are two girls playing in the grass. Giants looming over their toys, monstrously out of proportion. They're holding hands and spinning, leaning farther and farther back until their fingers rope together, chubby flesh and bone enmeshed. What do I see? Angular momentum around their center. A principle of physics in their motion. A girlish memory of grass-stained knees.
>
> I keep walking and recede from the girls' easy confidence in the world's mechanisms. I believe they exist, even if my knowledge of them can only be imperfect, a crude sketch of their billions of vibrating atoms. I believe this to be true . . .

I am on an orbit through the universe that crosses the paths of some girls, a teenager, a dog, an old woman . . . I could have written this book entirely differently, but then again, maybe this book is the only way it could be, and these are the only choices I could have made. This is me, an unreal composite, maybe part liar, maybe not free.

✳

TIPPETT: I sense that what you know about mathematics, and the kinds of ideas that you spend your life with, do leave you with a real nagging question about human freedom, about free will.

LEVIN: Absolutely.

TIPPETT: Talk to me about that.

LEVIN: I think it's a difficult question to understand what it means to have free will if we are completely determined by the laws of physics, and even if we're not. Because there are things—for instance, in quantum mechanics, which is the theory of physics on the highest energy scales—which imply that there is some kind of quantum randomness so that we're not completely determined. But randomness doesn't really help me either.

TIPPETT: It doesn't suggest to you that there is space for human decisions and for people to change the way things were built?

LEVIN: I don't see how it does. I don't see how it does.

TIPPETT: Okay.

LEVIN: You know, if something randomly falls in a certain way, how is that a gesture of will? So either will has to do with determinism—my will strictly determines an outcome—or it doesn't. It's very hard. There is no clear way of making sense of an idea of free will in a pinball game of strict determinism or in a game with elements of random chance thrown in. It doesn't mean that there isn't a free will. I've often said maybe someday we'll just discover something. I mean, quantum mechanics was a surprise. General relativity was a surprise. The idea of curved space-time. All of these great discoveries were great surprises, and we shouldn't decide ahead of time what is or isn't true. So it might be that this convincing feeling I have, that I am executing free will, is actually because I'm observing something that is there. I just can't understand how it's there. Or it's a total illusion. It's a very, very convincing illusion, but it's an illusion all the same.

TIPPETT: So for you, as a scientist, this convincing feeling, you simply can't take that as seriously as a calculation that you can prove no matter what?

LEVIN: Our convincing feeling is that time is absolute. Our convincing feeling is that there should be no limit to how fast you can travel. Our convincing feelings are based on our experiences because of the size that we are, literally, the speed at which we move, the fact that we evolved on a planet under a particular star. So our eyes, for instance, are at peak in their perception of yellow, which is the wave band the sun peaks at. It's not an accident that our perceptions and our physical environment are connected. We're limited, also, by that.

That makes our intuitions excellent for ordinary things, for ordinary life. That's how our brains evolved and our perceptions evolved, to respond to things like the Sun and the Earth and these scales. And if we were quantum particles, we would think quantum mechanics were totally intuitive. Things fluctuating in and out of existence, or not being certain of whether they're particles or waves—these kinds of strange things that come out of quantum theory—would seem absolutely natural.

What would seem really bizarre is the kind of rigid, clear-cut world that we live in. So I guess my answer would be that our intuitions are based on our minds, our minds are based on our neural structures, our neural structures evolved on a planet, under a sun, with very specific conditions. We reflect the physical world that we evolved from. It's not a miracle.

TIPPETT: As you have come to see things this way through your work as a scientist, do you live differently because of that? Do you raise your children differently or is it just a puzzle that you hold, that you carry forward?

LEVIN: The questions about free will? If I conclude that there is no free will, it doesn't mean that I should go run amok in the streets. I'm no more free to make that choice than I am to make any other choice. There's a practical notion of responsibility or civic free will that we uphold when we prosecute somebody, when we hold juries or when we pursue justice that I completely think is a practical notion that we should continue to pursue. It's not like I can choose to be irresponsible or responsible because I'm confused about free will.

TIPPETT: Okay.

LEVIN: That's being even more confused than me!

✳

The great mathematicians Janna Levin writes about in her novel, *A Madman Dreams of Turing Machines*, did not find the purity and clarity in life that they saw in logic. Kurt Gödel became delusional while at Princeton's Institute for Advanced Study. And for fear of poisoning, he starved himself slowly to death. Alan Turing helped invent modern computer science and was celebrated in England for helping crack Nazi codes during World War II. But he was later imprisoned and chemically castrated for admitting to a consensual homosexual affair.

He committed suicide in 1954 by eating an apple he had soaked in cyanide. In her novel, Levin writes, "One plus one will always be two. [Turing and Gödel's] broken lives are mere anecdotes in the margins of their discoveries. But then their discoveries are evidence of our purpose, and their lives are parables on free will. Against indifference, I want to tell their stories."

I asked Levin how the personal turmoil in these lives of logical brilliance informs her sense of purpose in individual lives and the universe.

✳

LEVIN: Well, I certainly think that both Turing and Gödel are examples of people living out their purpose. Even though they came to tragic ends, they were people who were committed, really, to meaningful pursuits. If you look at Turing, for instance, he was honest to the end. He really believed in being blunt and truthful. He couldn't pretend. He couldn't be a fake. He hated this idea of fakes and phonies. And he couldn't pretend to be somebody he wasn't. He couldn't pretend to be heterosexual even if it meant imprisonment or lethal poison. There is a person who, even though he might not have believed in free will, still behaved in a way that I think most people would hold up as being responsible, responsible for himself and believing in truth. And Gödel also, even though he went very astray in his compulsions and his paranoia and his imaginings, was very committed to being truthful, in a sense, to really following logic where it led him and to not deceiving himself or taking an easier path. So both are admirable examples of people living up to their innate purpose.

TIPPETT: And those are two extreme stories. I do want to say that although there is real tragedy in them, you present them in a very human light. We also see what was wonderful about these human beings and what they brought into the world. So I don't want to say that, you know, here are these stories just of tragedy.

LEVIN: Right.

TIPPETT: But a more mundane question is, how does the messiness of experience, of all of us, not just what we can know but how life unfolds, how does that impinge on the ultimate reality of what we can know and achieve through logic and through science?

LEVIN: I would argue that we should never turn away from what nature has to show us. We should never pretend we don't see it just because it's too difficult to confront it. That's something that I don't understand about other attitudes that want to disregard certain discoveries because they don't jell with their beliefs. One of the painful but beautiful things about being a scientist is being able to say, "It doesn't matter what I believe. I might believe that the universe is a certain age, but if I'm wrong, I'm wrong." There's something really thrilling about being committed to that. And so, in my own life, I don't feel that causes me problems. In a lot of ways, I've also made easier choices than my two heroes whom I wrote about.

TIPPETT: Right.

LEVIN: I have children. They did not have children. I have a certain physical comfort around me that they didn't have. In a way, I'm a much more connected person than either Gödel or Turing, though I still have some of the affinities that they had. Maybe that means that I'll never go as far as they went in my own discoveries. I hope that's not the case, but I can imagine maybe it will be. Maybe there is a tradeoff. Maybe sometimes you just have to abandon everything and pursue nothing but that. I'd like to think that if I'm lucky, I'll just get better at honing in on the jugular things, so that I can still make progress and discoveries as a scientist or have epiphanies as a writer. But I guess we all have to find that particular balance.

TIPPETT: I also sense that you're pursuing questions, beliefs, hunches about the meaning of life or just about what matters to you in a form that calculations simply can't contain or convey, that simply can't be captured in numbers.

LEVIN: You mean by writing a book, for instance?

TIPPETT: Yes, by writing a book.

LEVIN: Or being engaged with the arts.

TIPPETT: Right.

LEVIN: Well, that's true. I think that the answers that we're going to get, the discoveries that we're going to make, are going to be in mathematics. But they're going to be meaningless to us unless they're integrated into a human perspective where we understand why we ask the questions, what the significance of the answers is for us, and how the world is going to change as a result of having made those discoveries. That's why I can't quit one and become completely committed to the other.

TIPPETT: Right.

LEVIN: I continue to go back and forth between the two subjects.

TIPPETT: Your book about two scientists led me on this path of reading other biographies of scientists. So I've been reading James Gleick's biography of Newton, another very complicated character. And what Newton discovered wasn't just important, it absolutely changed the way people thought about the world. So I'm curious, what are you working on right now that is probably not accessible to most of us, where we wouldn't even know that these kinds of discussions are taking place? What are you working on that also starts to reshape the way you see the world around you and the way you move through it?

LEVIN: It's funny, people have often asked, when I've been describing the work that I'm doing, "Well, why should I care about that?" I'm talking about extra dimensions, and that maybe the universe isn't three-dimensional, but that maybe there are extra spatial dimensions. It is very abstract. We could do a whole show hammering that out.

TIPPETT: Yeah.

LEVIN: Why should you care about that? You know, our taxes are high. We're at war in Iraq. These are fair questions, but the notion of multidimensional space changes the world in such a fundamental way. We cannot begin to comprehend the consequences of living in a world after we know certain things about it. We cannot imagine the mindset of somebody pre-Copernicus, who thought that the Earth was the center of the universe, and that the Sun and all the celestial bodies orbited us.

It's really not that huge a discovery in retrospect. In retrospect, so we orbit around the Sun. We take this to be commonplace. And there are lots of planets in our solar system, and the Sun is just one star out of billions or hundreds of billions in our galaxy, and there are hundreds of billions of galaxies. We become little dust mites in the scheme of things. That shift is so colossal in terms of what it did, to our global culture, our worldview. I can't begin to draw simple lines to "This is what happened because of it" or, "That's what happened because of it."

TIPPETT: Right.

LEVIN: We see ourselves differently, and then we see the whole world differently. And we begin to think about meaning—and all of these questions that you've brought up—completely differently than we did before. I'd feel the same way if we discovered that the universe is finite or if we discovered that there are additional spatial dimensions. These things will impact us in ways that we can't just draw simple cause-and-effect arrows to.

TIPPETT: Does it make you react to simple things differently in your life, because you are closer to that cutting edge of knowledge right now?

LEVIN: Well, I will often look at what people feel is very important and not identify with what they think is very important. I have a hard time becoming obsessed with internal social norms, how you're supposed to dress or wear your tie or who's supposed to . . . for me, it's so absurd because it's so small. It's this funny thing that this one species is acting out on this tiny planet in this huge, vast cosmos. So it is sometimes hard for me to participate in certain values that other people have. I guess there is a shift of what I think is significant and what I think isn't. And if I try to look at that closely, I would say that things totally constructed by human beings I have a hard time taking seriously. And things that seem to be natural phenomena, that happen universally, I take more seriously, as more significant.

TIPPETT: Give me an example. I mean, sometimes it's hard to draw the line. Give me an example of something for you that would be totally humanly constructed.

LEVIN: Well, let's say . . .

TIPPETT: Aside from dress codes.

LEVIN: Right. Actually, this is going to sound really dangerous, but even things like who we elect as an official in our government. Of course, I take our voting process seriously and I try to be politically conscious. But sometimes, when I think about it, I have to laugh that we're all just agreeing to respect this agreement that this person has been elected for something. That is really a totally human construct. We could turn around tomorrow and all choose to behave differently. We're animals that organize in a certain way. It's not that I completely dismiss it, but I think a lot of the things we are acting out are animalistic, consequences of our instincts. They aren't, in some sense, as meaningful to me as the things that will live on after our species comes and goes. Does that make any sense?

TIPPETT: It does—it makes a lot of sense. It's perspective that you bring, that you have that's different, that's a bit larger, that's cosmic.

LEVIN: And it doesn't mean that I'm dismissing things as unimportant. I'm really pained by what's going on in the world. But my perspective is to look on it as as animals acting out ruthless instincts and unable to control themselves—even though other people think that they're being very heady and intellectual.

TIPPETT: So I do believe, I think I know, that something deep is met in human beings in a sense of being part of something larger than oneself, being part of something big.

LEVIN: Well, we are a part of something larger than ourselves. We definitely are made up of material that was synthesized in the cores of stars, a previous generation of stars. We come from a very specific series of events in this universe. If they hadn't happened, we wouldn't be here.

TIPPETT: Some people might listen to this and feel that if you really internalize this, that possibly everything is predetermined, that we in fact are not free in any way, that we are behaving like animals even when we think we're at our most civilized, that life would somehow be robbed of joy and hope and transcendence. I don't experience you as a person without joy, hope, and transcendence.

LEVIN: No, I don't feel that way at all. I have a fifteen-month-old daughter and a four-year-old son. And the overwhelming feelings I have for them, even if I believe that they're instinct, do not fade one bit because of that. It matters to me not at all that I have evolved to feel that way. It doesn't take anything away whatsoever. That feeling is as real, as strong, as beautiful, as meaningful as it is for somebody who believes otherwise. I've never really understood the argument that it takes the shine off of things.

For instance, let's say somebody had a belief system in which it was simply posited that carbon came out of, I don't know, a blue sky one day. That wouldn't make me feel any more meaning about who I was in the world. It feels much richer to me to imagine that a cold, empty cosmos collapses with stars, and stars burn and shine, and they make carbon in their cores and then they throw them out again. And that carbon collects and forms another planet and another star and then amino acids evolve and then human beings arise. That, to me, is a really beautiful narrative.

✳

Here's another passage from Janna Levin's book *A Madman Dreams of Turing Machines*:

> They are here in our minds, Turing's luminescent gems, Gödel's platonic forms. There are no social hierarchies to scale. No racial barriers. Given to us along with our brains. Built into the structure of our thoughts—no bullying into blind faith, no threats of eternal damnation—just honesty, truth, and reason.
>
> I am here in the middle of an unfinished story. I used to believe that one day I would come to some kind of conclusion, some calming resolution, and the restlessness would end. But that will never happen. Even now, I'm moving toward a train. My heart is thumping. My lungs are working. There is a man, a woman, a bench, the glasses, the smooth hair and umbrella. We are all caught in the stream of a complicated legacy—a proof of the limits of human reason, a proof of our boundlessness. A

declaration that we were down here on this crowded, lonely planet. A declaration that we mattered, we living clumps of ash, that each of us was once somebody, that we strove for what we could never have, that we could admit as much. That was us—funny and lousy and great all at once.

✳

TIPPETT: It seems to me that there is so much beguiling mystery in science right now. Even language, like dark matter.

LEVIN: We can be pretty corny, too, you know? All kinds of acronyms and . . .

TIPPETT: Right. But whether that's the same way religious people talk about mystery or not, there's real mystery in it. Isn't that right?

LEVIN: Yeah. I think the secret you are uncovering is that scientists often share a very childlike wonder for the world. A lot of the language that we invent about the universe reflects that kind of childlike experience. There is really that feeling of excitement over learning about the universe, and wanting it to sound a certain way. Wanting the language to reflect the mystery and the magnitude of what we're learning. I think that's what you're picking up on.

TIPPETT: I know that you're now working on the idea of whether the universe is infinite or finite. And somewhat against the grain, you are pondering whether the universe is finite. Explain that to me.

LEVIN: There are a handful of people who started getting interested in this around the world several years ago. It's similar to the idea of the Earth. If you're standing, as I am, in New York City and you walk in a straight line, and then you swim in a straight line, and then you walk again and swim again, you keep going in a straight line as far as you possibly can go, you will end up coming back to New York City because the Earth . . .

TIPPETT: Okay.

LEVIN: . . . is not infinite, even though it doesn't have an edge off of which you would just sort of fall. So in space-time, it might be something like that. I travel in a rocket ship and I find myself coming back to where I started. I think I left the Earth behind me, I see it go away behind me. And as I approach some planet in front of me, I realize, "Whoa, that's the Earth again."

TIPPETT: You've made this interesting observation that several times in history science has acknowledged limits, right? You'd be putting finitude to infinity, and that in fact has made great leaps forward possible.

LEVIN: Yes, it doesn't mean that we throw up our hands and say we can't know anything. Mathematics has limits, we don't do mathematics anymore; where the speed of light is a fundamental limit, we stop doing physics. It's really been exactly the opposite. Mathematics has limits, and that leads people to invent a computer. The speed of light has a finite limit, which is what Einstein proposed, and he invents special relativity, and eventually a theory of curved space-time based on this observation.

So it opens up this huge way of thinking about the world, when we accept our limits and just move on. Quantum mechanics was the other example. Quantum mechanics implies a fundamental uncertainty in what we can know about physical reality. And by accepting this, we make these enormous discoveries. Einstein said this funny thing, that only two things are infinite, the universe and human stupidity. And then he said, "I'm not so sure about the universe."

He knew that it was conceivable that the universe wasn't infinite, but he wasn't sure how to go about it. Only later did we understand how to actually handle it. If we were to discover that the universe was finite, it would again be something like what happened with Copernicus or like understanding that there was a Big Bang. It's hard for us to remember what it was like before the discovery of the Big Bang itself. That's just such a part of our worldview now.

TIPPETT: That there was a beginning point.

LEVIN: That there was a beginning, that the universe hasn't always been here, that it isn't permanent, and unending, and unalterable.

TIPPETT: We spoke at the very beginning about Kurt Gödel, one of the two scientists you wrote about in your novel. So he said there are things that are true that mathematics cannot prove. They might still be true, but the idea was you would have to go outside mathematics to know that. And you use phrases in your writing like, "We can't see the logic of them until we step outside the logical framework." You say something like, "We have to look at them out of the corner of our eye." To me, that again seems so resonant with life as I know it. And I wonder if that's a kind of idea that you also find you can translate into other aspects of knowledge and experience.

LEVIN: I definitely think it's the reason the book was structured as a novel. I tried to stick as close to fact as possible. It's not the facts that I'm changing, it's the approach to the facts. And it's a sort of confession that no matter how I list these facts, I am somehow not able to get at the truth. The truth doesn't drop out like a theorem if I follow certain logical steps. Maybe it's saying something also about maybe my own approach to science.

No matter how much I follow these logical steps, no matter how much I make real discoveries that will be unambiguous, I hope my approach to the truth, in the bigger sense of the meaning of the word, will always be a little bit out of the corner of my eye—the visceral experience of what it really means or what the implications are. There are no true things—except for things as crisp as one plus one equals two—that are unambiguously true.

And yet we know we're getting closer to the truth even though we can't always prove it.

Here, in closing, is a reading from the final chapter of Janna Levin's novel, *A Madman Dreams of Turing Machines*:

Here I am, in New York City. It is the twenty-first century. This is as good a place, this time as good a time as any.

I am stepping off the curb. The subway entrance is just across the street. Big green orbs signify that the downtown entrance is open. Artificial light competing against the sun. There are children in the park; a woman with silver hair and a long, old coat; a couple on a bench. A yellow plague of taxis infests the streets. I walk behind some and in front of others. We all move along fixed trajectories, following our prescribed arcs. The plague disappears behind the stone wall as the old steps take me down into the subway . . .

There is no ending. I've tried to invent one but it was a lie and I don't want to be a liar. This story will end where it began, in the middle. A triangle or a circle. A closed loop with three points. A wayfaring chronicle searching for a treasure buried in the woods, on the streets, in books, on empty trains. Craving an amulet, a jewel, a reason, a purpose, a truth. I can almost see it on the periphery, just where they said it would be, glistening at me from the far edges of every angle I search.

7

*

Science That Liberates Us from Reductive Analyses

"GETTING REVENGE AND FORGIVENESS"

We did not plan to put the subject of revenge and forgiveness on our schedule for the weekend after the 2008 election—it just landed there. But it did seem right, and good, and helpful, with a decidedly real-world vigor and clear sightedness.

I'd been intrigued by what I knew of Michael McCullough's research, and I was hooked by this line at the beginning of his book *Beyond Revenge*: "I wrote this book for people who want to bypass all of the pious-sounding statements about the power of forgiveness, and all of the fruitless sermonizing about the destructiveness of revenge. It's for people who want to see human nature for what it really is." Part of my passion for the spiritual and religious aspect of life is my delight in the fact that here we dwell solemnly not only on God but on what is ordinary and human; we attempt to see human nature for what it really is, and find meaning and mystery right there.

I first began to gain a solemnity about the revenge impulse in human life when we worked, in the early days of *Speaking of Faith*, on a show about the death penalty. I came to understand that

172 * EINSTEIN'S GOD

revenge is the original "criminal justice system." For most of human history, before the rule of law, before structures of justice that transcend the messiness of human interaction, the threat of retaliation has been a primary tool humans possess to pursue justice and also to regulate cycles of violence. The ancient "eye for an eye" teaching of the Hebrew Bible—which is often cited as a justification for extreme revenge—arose in this context. It was not designed to champion extreme punishment, but to limit revenge in terms of equity and fairness—as in, "you may not slaughter the entire family of the person who harmed you or your loved one; you may only take an eye for an eye."

And now, as Michael McCullough lays out expertly and passionately, science is able to document how normal, and in a sense, how sensible our instinct for revenge is. It has served a purpose in human life and in the primate world. We are hardwired for what looks in the brain like a "craving" for revenge, a felt need that begs for satiation. And though we do range in this conversation into the realms of global geopolitics and murderous revenge on a societal scale, Michael McCullough is more interested perhaps in the mundane forms this craving takes: in our interactions with obnoxious neighbors and irritating coworkers or, for example, the political candidates we oppose. He notes that Americans have a tendency to see revenge as a mark of cultures more primitive than their own. But he points out, provocatively, that between 1974 and 2000, 61 percent of all school shootings in the United States had revenge—often for bullying—as a trigger.

Here is the good news: science is also revealing how forgiveness, like revenge, is hardwired in us—it is purposeful and normal. We tolerate and excuse the deficits and mistakes of those we know and love and work with—and even those we don't love but need to work with—a hundred times a day without ever glorifying these moments with the lofty word "forgiveness." School shootings, ethnic slaughter, and road rage garner headlines, skewing our sense of our collective character. However, McCullough says, forgiveness doesn't work in real life as it too often works in media portrayals of dramatic stories of conversion and high emotion. Actually, we forgive in good part because it is in the interests of our genetic pool to do so. The evolutionary payoff for the forgiveness of mistakes by people we are close to or whose work we depend upon, for example, is survival. Michael McCullough says to think of forgiveness as a trait of the weak and the vulnerable reflects a simplistic imagination about human nature and evolutionary biology. And he has the science to give us a more complex imagination about both.

This is science, in other words, that liberates us from reductive analyses of human nature—that is to say, of ourselves and those around us. If we accept the normalcy of our instincts both to revenge and forgive, we have more control over both. Among the practical tools McCullough offers for moving forward, in this way, here is one of the most simple and challenging. we embolden the forgiveness instinct when we come to see others as having value. In this light, religious traditions have more than straight teachings on forgiveness to offer our culture. Perhaps more practically, they have rich, ancient, cross-generational resources for seeing, knowing, and honoring the dignity of "the other," whether enemy or friend, neighbor or stranger.

On the cautionary side of McCullough's insight, there is a realization that under the right conditions, we are all vulnerable to falling back on revenge as a form of justice. This helps explain the fact that sectarian cycles of revenge often erupt after the fall of dictatorships, like that of the former Yugoslavia or that of Saddam Hussein. Such regimes take all the revenge function on themselves and keep normal human dynamics artificially in check. McCullough's science makes a sobering case for the necessity of the basic rule of law—in Iraq or in an American inner city—if human beings are to live up to their moral potential. The need to understand the instincts for revenge and forgiveness, and to govern them, may be attaining a new urgency in a globalized world, and in the wake of globalized financial crisis. I know that Michael McCullough's analysis has been ringing in my ears—anchoring both my concerns and my hopes—as I've watched that unfold, and as I consider the ongoing challenge of laying American "culture wars" to rest.

＊

Getting Revenge and Forgiveness

KRISTA TIPPETT, host
MICHAEL MCCULLOUGH, psychologist and author

＊

*M*ichael McCullough is a professor of psychology at the University of Miami in Coral Gables, Florida, where he directs the Laboratory for Social and Clinical Psychology and also teaches in the Department of Religious Studies. For his book, *Beyond Revenge*, he analyzed extensive data from social scientific studies on humans and animals as well as biology and brain chemistry. I'll discuss with McCullough what he is learning about forgiveness. But he stresses that to reimagine the human capacity for forgiveness, we must first challenge our ideas about the human inclination to seek revenge.

Western religious and therapeutic mindsets have come to imagine revenge as a disease that can be cured by civilization. It hasn't been seen as a natural, biologically driven impulse to which we all remain prone under certain circumstances. And at the same time, the seemingly colder eye of evolutionary biology has analyzed ruthlessness as an advantage in the relentless arc of the survival of the fittest. Forgiveness in both of these scenarios is a rare transcendent quality, a cure for revenge, albeit one that would never help human beings really triumph.

Michael McCullough says this view of the world is based on simplistic understandings of both human nature and evolution.

TIPPETT: One of the things you seem to be talking about is reclaiming the normalcy of both revenge and forgiveness as a part of human nature. I'd like to talk about revenge first, if we could— why revenge is in us and what purpose it has served even in evolutionary terms.

McCULLOUGH: Here's what you see all throughout the animal kingdom—and this is where I really got interested. One study that really got my attention was a study on chimpanzees, which showed that if a chimpanzee is harmed by an individual that it's living with, it has the ability to remember who that individual is and target aggression back at that individual in the ten minutes, twenty minutes, hour later. I was surprised to know that chimpanzees had these kinds of mental abilities. I had to learn more. I wanted to know where else you see this in the animal kingdom. It turns out that you see it in other kinds of primates, such as one type of monkey that I like a lot, a monkey called the Japanese macaque. Japanese macaques are very status-conscious individuals. They're very intimidated by power; let's just put it that way. So if you're a high-ranking Japanese macaque and you harm a lower-ranking Japanese macaque, that low-ranking individual is not going to harm you back. It's just too intimidating. It's too anxiety provoking. What they do instead, and this still astonishes me, is they will find a relative of that high-ranking individual and go seek that low-ranking cousin or nephew out and harm him in retaliation.

TIPPETT: That does sound like human behavior, doesn't it?

McCULLOUGH: Right. And here's the kicker: when they're harming this nephew, most of the time they're doing it while the high-ranking individual is watching. They want the high-ranking individual to know that I know you're more powerful than I am. But rest assured, I know how to get at what you care about and what you value.

TIPPETT: I had this realization a few years ago when we did a program on the death penalty. It might seem so simple but it seemed so stunning to me to realize that the criminal justice system, and especially the death penalty in history, progressed because before there was any kind of criminal justice system, human societies regulated themselves by precisely that kind of revenge you're describing.

McCULLOUGH: Throughout most of human history we have not lived in complex societies with governments and states and law enforcement and prisons and contracts we could enforce in a court to get people to do what they agreed to do. The mechanism that individuals relied upon to protect themselves and to protect their loved ones and to protect their property was fear of retaliation. If they could broadcast that fear of retaliation to the individuals they lived with, to their neighbors, to the people on the other side of the hill—if you could cultivate a reputation as a hothead so people knew not to mess with you—that was like an insurance policy. You're absolutely right that in a lot of the world this is still going on. And any time you disrupt that system—that system of government, that system of policing, that system of law enforcement—so people can't trust that their interests are going to be protected, that desire for revenge comes back. People will take revenge back into their own hands to protect themselves.

TIPPETT: I think you're also saying in your research that in terms of what we know about the brain—the emotions, the reactions, that arise in response to grievance—we are hardwired to have those reactions. They serve a purpose. I remember Sister Helen Prejean saying to me when we did that work on the death penalty—and she's a great opponent of the death penalty—she said, "Anger is a moral response."

McCULLOUGH: That's right. It certainly is. Anger in response to injustice is as reliable a human emotional response as happiness is to winning the lottery, or grief is to losing a loved one. And if you look at the brain of somebody who has just been harmed by someone—they've been ridiculed or harassed or insulted—we can put those people into technology that allows us to see what their brains are doing. We can look at what your brain looks like on revenge. It looks exactly like the brain of somebody who is thirsty and is just about to get a sweet drink or is hungry and about to get a piece of chocolate to eat.

TIPPETT: It's like the satisfaction of a craving?

McCULLOUGH: It is exactly like that. It is literally a craving. What you see is high activation in the brain's reward system. So, again, this is one of the messages it's important for me to try to get across. The desire for revenge does not come from some sick dark part of how our minds operate. It is a craving to solve a problem and accomplish a goal.

TIPPETT: And then what is especially intriguing about your work as well, and perhaps even more surprising than the fact that revenge is natural, is that you are really suggesting also from a scientific perspective that we have a forgiveness instinct, an aptitude for forgiveness. And that has been crafted by natural selection just like revenge.

MCCULLOUGH: I expected to find, frankly, less research as I dug through hundreds of scientific articles on the naturalness of forgiveness, but, boy, was I wrong. As it turns out, a lot of biologists have been trying to figure out what allows human beings to be the cooperative creatures that we are. We're cooperative with each other in a way that really makes us pretty unique among mammals. We cooperate with our relatives, but lots of animals do that. We go further and we cooperate with people we've never met. We cooperate with people we're not related to. And by virtue of our ability to cooperate with each other, we can build magnificent cities and radio stations and do all kinds of wonderful things. But one of the ingredients you have to have to get individuals to cooperate with each other is a tolerance for mistakes.

TIPPETT: Interesting.

MCCULLOUGH: You can't get organisms willing to hang in there with each other through thick and thin and make good things happen, despite the roadblocks and the bumps along the way, if they aren't willing to tolerate each other's mistakes. Sometimes I'm going to let you down. And maybe it's not even intentional, but I'm going to get distracted and I'm going to make a mistake. And if you take each of those mistakes as the last word about my cooperative disposition, you might just give up and so no cooperation gets done. So, really, our ability to cooperate with each other and make things happen that we can't do on our own is undergirded by an ability to forgive each other for occasional defects and mistakes.

TIPPETT: Here's a passage from your book—and again, a lot of this seems so basic when you articulate it. You've said that everyday acts of forgiveness are incredibly common among people who know each other. We think of forgiveness as heroic acts and there are always heroic examples of forgiveness. But you said we think of it as this "balm for a wound. Yet in daily life, forgiveness is more often like a Band-Aid on a scrape, and at first glance, perhaps only slightly more interesting. But of course uninteresting doesn't mean unimportant."

MCCULLOUGH: Right. And this again was part of my attempt to do violence, I guess, to this metaphor of forgiveness as this difficult thing that we have to consciously practice and learn, because we don't know how to do it on our own. I forgive my seven-year-old son every day. Right?

TIPPETT: Right.

MCCULLOUGH: He's an active, inquisitive seven-year-old who sometimes accidentally elbows me in the mouth when we're cuddling and sometimes puts crayons on the walls. And yet it seems demeaning to call it forgiveness.

TIPPETT: To even call it forgiveness.

MCCULLOUGH: I wouldn't dignify it with the term "forgiveness." It's just what you do with your children. You accept their limitations and you move on. He broke my tooth once when I was drinking out of a water glass. Parents have a million of these stories. But you don't put any effort into forgiving. It naturally happens and you move on. And there's a great evolutionary story about why it comes so easy in those kinds of circumstances, too.

TIPPETT: Which is pretty obvious, I guess.

McCULLOUGH: Evolution wasn't kind to individuals who would seek revenge against their genetic relatives, bottom line, right? So we have this natural tolerance for the misbehavior of our children. At that level it is incredibly mundane. We would never even give it a second thought. And yet we do it over and over again.

✳

One of the most high-profile figures of public forgiveness in the United States in recent years was Bud Welch. His twenty-three-year-old daughter, Julie, died in the bombing of the Murrah Federal Building in Oklahoma City in April 1995. Here is a statement Bud Welch made before the 2001 execution of Timothy McVeigh, the terrorist responsible for the bombing:

> The first month after the bombing, I didn't want Tim McVeigh and Terry Nichols to even have trials. I simply wanted them fried. And then I finally come to realize that the reason that Julie and 167 others were dead is because of vengeance and rage. When we take him out of his cage to kill him, it's going to be the same thing. We will keep the circle of violence going. Number 169 dead is not going to help the family members of the first 168.

✳

TIPPETT: You do talk about some amazing examples of forgiveness, of public forgiveness, one of them being Bud Welch. But I sometimes think those kinds of examples that do make the news, like the bombing, also exalt forgiveness as something that's really beyond the reach of most of us most of the time. We hope we would be that gracious, perhaps, but it almost feels superhuman.

McCULLOUGH: Right. And if you look at Bud Welch and you look at that story from the outside and you ask yourself how can this man whose daughter was killed in this terrible explosion ever get over his rage, from the outside we have a really hard time imagining that. But if you look at the story of Bud Welch, actually what you find is he had a lot of help along the way. And if you look at the story very carefully, you can learn a lot about how the human mind evolved to forgive and what kind of conditions activate that instinct in human minds. A lot of those conditions ended up falling into place for Bud. In fact, he doesn't talk about forgiveness in that case as having been some massive struggle.

TIPPETT: It was incremental, wasn't it? It gets reported as an act, but in fact it was a process.

McCULLOUGH: That's right. And along the way, there were events he actually made happen for himself that made forgiveness easier. He sought out Timothy McVeigh's father and visited him one day at the McVeigh home. He had this moment he describes, when he saw Timothy's picture on the mantel. It was a high school graduation picture. They were just making small talk. And Bud said to McVeigh's father, "God, that's a good-looking kid." And the tears just began pouring out of the elder McVeigh. He realized then that here was another father on the verge of losing a son, of losing a child. And that immediate experience of sympathy and compassion went a tremendous way in facilitating the forgiveness process for Bud.

So right off the bat, this real human interaction starts to turn forgiveness from something difficult to do to something that's easier to do.

TIPPETT: So this is getting to one of the really important points I think you make with your work: that if we can understand this forgiveness instinct, even in terms of evolution, we can start to create conditions where it can be empowered.

McCULLOUGH: The first is safety. Human beings are naturally prone to forgive individuals they feel safe around. So if we have an offender apologizing in a way that seems heartfelt and convincing and has really convinced us that he can't and won't harm us in the same way again, that's a point for forgiveness. Again, the human mind evolved for forgiveness to be something worth its while. And any successful organism is unlikely to have a mechanism in it that says, "Just keep stepping on my neck. It's okay."

But if you can convince me that you're safe, that I don't have to worry about being harmed in the same way a second time, maybe I'm willing to move a little bit forward.

TIPPETT: But it seems like that would be the hardest condition or assumption to put in place in the context of many of the worst cycles of revenge in our world.

McCULLOUGH: Sometimes safety comes through things like the rule of law. Sometimes safety comes through you as a small-business owner dusting off that employee manual and asking yourself, what is in here that would instruct an employee on what to do if they were being systematically harassed by a co-worker? And what would insure that if there was a serious infraction, it would be dealt with in a way that restored that employee's sense of safety?

What can you do in your associations? Your condo association, say, when somebody has a grievance, when the neighbor has hired a band for a party at 12:30 on a Friday night. You need to know how to make sure that doesn't happen a second time, so that you don't then have to say, "I'm going to get back at that guy myself. I'm going to leave my garbage cans out all weekend long, which I know he hates."

TIPPETT: You're talking about revenge in ordinary life, whereas I think we're more comfortable talking about it in terms of warring tribes across the globe.

McCULLOUGH: The thing I like about these principles is that they're scalable. Usually, when people ask me about the book, they're actually less interested in the geopolitical stuff.

TIPPETT: Well, we'll get there. So what's the second condition we can create to make forgiveness easier, after safety?

McCULLOUGH: Value. We are inclined to forgive individuals who are likely to have benefit for us in the future. We find it relatively easy, as I was saying, to forgive our loved ones or forgive our friends or forgive our neighbors or our business partners, because there's something in it for us in the future. And the costs sometimes of destroying a relationship that's been damaged are just too high, because establishing a new one is so difficult to do. So relationships with value in them are ones in which we're naturally prone to forgive.

Michael McCullough observes that Americans have a tendency to ⸱e revenge as a mark of cultures more primitive than their own.

But he points out, provocatively, that between 1974 and 2000, 61 percent of all school shootings in the United States had revenge, often for bullying, as some kind of trigger. His perspective also helps illuminate why partisan rancor seems to spin out of proportion during an election cycle, as political campaigns accentuate the differences between candidates and constituencies.

✳

MCCULLOUGH: We tend to view other people who hold positions different from ours as having much more similarity to each other than we do among ourselves. We can see the great variety in our own positions . . .

TIPPETT: But we can't see the variety in other people's positions?

MCCULLOUGH: That's right.

TIPPETT: That's interesting.

MCCULLOUGH: We tend to paint them with the same brush. And so we tend to really simplify positions that other groups have or people on other sides of positions. We tend to actually view them as more partisan and more extreme on average than the average really seems to be. There's something about how the mind works and how it processes groups. When we think about people from over there, that other group, we don't really view them with the same sort of humanity that we afford our own groups. If you think about an issue that you feel strongly about, and that you know a lot about, you can see that actually there are a lot of people who have different views, that they're not exactly the same, and that allows you to view them as human beings.

TIPPETT: Right.

McCULLOUGH: But perhaps because of how the mind was actually designed to work, we have a harder time affording that kind of benefit of the doubt to other groups. So if we know that, then . . .

TIPPETT: If we know that about ourselves, if we can get an awareness about that, perhaps that is a beginning.

McCULLOUGH: Then you can begin to say, "Well, they're just a group of human beings, too, trying to muddle their way through to a position that's going to work for them." And maybe that kind of recognition of their diversity can help. Then maybe we'll have less anxiety about interacting in a civil way.

TIPPETT: Let's go to the geopolitical level, where you have cycles, generations, of grievance and revenge layered on top of each other. And yet you tell stories and I've met amazing people who have reached out to people on the other side, have come to see the other group as human and have formed friendships. Those kinds of stories don't tend to be in the headlines. We hear the headlines of continued violence and continued animosity instead.

McCULLOUGH: Right.

TIPPETT: But from the studies you've seen and from what you know about how these things play themselves out in different societies, is it possible even, say, in the Israeli-Palestinian crisis, that one day those networks of forgiveness would reach such a critical mass that the balance of the entire political and inter-societal dynamic is shifted? How does that kind of collective change really happen?

MCCULLOUGH: Some of it happens when people become too tired to fight.

TIPPETT: You tell a story from northern Uganda, where you say an epidemic of forgiveness has grown out of fatigue as much as anything else.

MCCULLOUGH: That's right. Sometimes the costs of maintaining grievances are so high that individuals and their groups will decide that they've pushed themselves to the brink. They've demonstrated their insistence on defending themselves, and they've shown that they will defend themselves to the end. Having done that, it becomes possible to try to find a new way.

So Uganda has been at war for many years. And part of the strategy of one of the rebel groups—it's a group called the Lord's Resistance Army, headed by a man named Joseph Kony—part of their strategy has been to abduct children, boys and girls, from their villages and from their tribes and take them off into the woods and essentially brainwash them.

TIPPETT: They've also had the children do horrible things before they leave. Killing their own siblings, so that they can't go back, they're so ashamed that their parents won't take them back. It's terrible.

MCCULLOUGH: Yes. They send them back to kill their own families, their own villages, their own tribes, to maim them, to disfigure people unrecognizably, to cut off their lips and ears and noses. They give the girls as child brides to the soldiers. And through this really heartless, brutal tactic, they do a couple of things. One is that they destroy the culture of these villages, the fabric of their own history. And they also create new foot soldiers for their army. The costs of this have been so high both from a security point of view and from a cultural point of view. And many of the rank and file, just regular people, particularly this one group called the Acholi, have just simply grown so tired of these cycles of violence and their inability to solve them using military force that they've really been pressing the government to offer official amnesty. Not only to Kony, but to any of the children, any of the sons and daughters of their own villages who've been spirited away like this and brainwashed and turned into killers. They've used radio broadcasts, word of mouth, newspapers, any vehicle they can get hold of to send this message out: If you will come back to your village, lay down your arms, meet with the elders, meet with the community, and work out a plan for demonstrating your desire to rejoin us, we'll let you rejoin us as a member of our community in good standing.

TIPPETT: That's pretty amazing, isn't it?

McCULLOUGH: And they've been coming back in groups as large as 300, 400, 500, 900. Laying down their guns, working out plans for reparation. Trying to find some way to compensate victims for the harms they've caused. They're doing this at risk to themselves, mind you. These returnees now have to worry about the villagers' own desires for revenge against them. They take a risk in coming back and yet many of them are doing it, in part because there just isn't another way.

✳

Here is a Welsh ethnomusicologist, Peter Cooke, describing how the Ugandans Michael McCullough mentioned have integrated grief, outrage, and a longing for forgiveness in the music they sing. And their songs explicitly address warlord Joseph Kony:

They live in slums around Kampala. What do they sing about? First of all, they preserve some of the songs from their village competitions. "We are number one. We are the best group. We are going to win," this kind of thing. Secondly, they'll sing about this war in the north, how awful it is. But in the same song, when they're complaining that their women are raped, that their sisters give birth in the bush, they will say, "Kony, come talk with us. Come talk with us. Let's get it settled." "Otti"—who's now dead, by the way, killed by Kony—"come talk with us. It's time for peace." And these songs are being sung at the same time as bureaucracy overseas will say the International Criminal Court wants to arrest Kony and try him. There's a lot of forgiveness for the sake of a lasting peace and building one as soon as possible.

Michael McCullough's explanations of the biological basis of forgiveness are revealing in light of conversations I've had across the years on human dynamics below the surface of the Israeli-Palestinian crisis, for example, and South Africa's Truth and Reconciliation Commission. I asked McCullough how his research might inform the way we think about possible outcomes of sectarian violence in Iraq.

✳

McCullough: We replaced one of the truly awful dictators of the late twentieth century when we removed Saddam Hussein. You know the story there. And yet it is also true that when we did that, and particularly when we disbanded the army, we did away with the only structure capable of holding a lot of very old tribal and ethnic and sectarian grudges in check.

TIPPETT: This is a really interesting point you make, that the effect of strong governments cuts both directions. Repressive governments squelch or kind of take on all the revenge impulses within the society.

McCULLOUGH: That's right.

TIPPETT: So that helps me understand why sometimes when you have terrible regimes fall apart—the Soviet Union or Saddam Hussein's regime and even in South Africa—some old rivalries come to the surface.

McCULLOUGH: That's right. I like to ask people to look out their windows of their office or their homes. And imagine what your life would look like if the police and the National Guard and the fire department and the paramedics stopped working tomorrow, because of a natural disaster. People are hungry. People have needs. How would you put security into place yourself? You would probably find your friends and find your family and you'd circle the wagons.

TIPPETT: So in terms of what average people can do in the course of more ordinary lives, let me ask you the question this way. How do you conduct yourself differently with people you fundamentally disagree with on important social issues, with irritating people at work? How do you conduct yourself differently because of what you know scientifically in this research you do?

McCULLOUGH: The thing I have realized is that many times if you've been harmed by somebody, you don't have any choice but to try to forgive on your own. Because the person's gone, the person's dead, the person will have nothing to do with you.

TIPPETT: Right.

McCULLOUGH: There's just no bridge there. But in lots and lots of cases, forgiveness is just a conversation away. So many people remember a hurt from junior high or high school, but you often find that there was never any conversation with that person who harmed them. And the conclusion I've come to is in many cases, if you want to forgive or if you want to be forgiven, you need to go out there and get it for yourself. And the way you go out and get it for yourself is by trying to have the kind of conversation that you want to have with the person you hurt. In my family, we apologize about a lot.

TIPPETT: Apology is an important concept for you. You say that it really, even biologically, is important for us.

McCULLOUGH: Apology is really important, because when I apologize to you for something I've done, you see me squirming. You see me uncomfortable. You see me trying to reassure you that I'm not going to harm you in the same way again. You see me giving you respect as a human being with feelings. And all of a sudden, I've turned on a lot of the slider switches that make forgiveness happen in your head.

TIPPETT: It's the next best thing to revenge.

McCULLOUGH: That's right.

TIPPETT: It's fulfilled some of those needs we have.

McCULLOUGH: There are so many people who, once they see someone who's harmed them cry and experience shame and experience humiliation for the way they've behaved—suddenly it's the forgiver who's doing the healing, who's reaching out to the perpetrator. This happens so many times. All people often need is this kind of vigorous conversation about the past. Now, if this were so easy, people would be doing it all day.

TIPPETT: Right.

McCULLOUGH: I don't pretend that. But at the same time, I really think we can't lose sight of the value of getting in each other's business a little bit and getting in each other's lives a little bit and being willing to try to make things a little bit uncomfortable and a little bit messy in the service of making them better.

TIPPETT: Again, if you just read the headlines, you read about what's going wrong in the world today, the worst and most entrenched crises. I do sense from your research that when you take a global view you feel there is progress, that on balance there's more reconciliation happening now. Is that right?

McCULLOUGH: I'm so optimistic about our future. Because, again, if you look at that long arc of history, as you suggest, what you see is—for example, the homicide rate. We worry about the homicide rate, as we should. It goes up some years, it goes down other years, and we worry. But over the long arc of history— take western Europe—homicide rates are a twentieth and in some countries a fiftieth of what they were six hundred, eight hundred years ago. Right? So if we take this long perspective, actually we're getting better and better control over human beings' potential for aggressiveness. And a lot of that homicide six hundred to eight hundred years ago was in fact vengeance motivated. But when we get control over those instincts and give people other tools to deal with their grievances, they will restrain themselves.

So Iraq may look dismal. It's been terrible for our country and the world in so many ways. And yet I see coming out of it, whenever that is, a society that's going to rebuild itself into a peaceful society. I don't know how long it will take—it's above my pay grade, as they say—but this is what societies tend to do. They tend to find the best way to rebuild in the aftermath of these kinds of collapses in ways that will promote cooperation.

TIPPETT: And you're saying that on the basis of lots of research, aren't you? This is not just wishful thinking.

McCULLOUGH: If you put societal structures in place where people feel their rights are protected, and they see a way forward for making a living in a peaceful way, and there's security, they prefer peace over war, every time.

TIPPETT: So from everything you know, what feels really important for you to pass on to your children, practically?

McCULLOUGH: I have a four-year-old daughter who's a little bit too young for this still. But with our seven-year-old, I really have tried to encourage him to be vigorous about acknowledging his mistakes and the harms that he causes his friends, whether that's just a careless word or excluding somebody from a game. Because so much of forgiveness comes down to interaction. It comes down to knowing that an offender is not the person you thought he was when he hurt you, or she was when she hurt you. It's changing that perception. It's simple things. We try to teach him what someone needs after they've had their feelings hurt. We think if we can explain to him what the mind needs after someone's been offended, then we can teach him how to be vigorous and not worry about having to look like he's right all the time or having to look like he's perfect or denying his mistakes. If he can own up to them, that's a vigorous healthy way to keep his friendships intact.

TIPPETT: When I look at all your research and have this conversation with you, it seems to me that religion can play a constructive role with teachings about caring for the other.

McCULLOUGH: Absolutely. One of the best things we can do with religious faith is give people an appetite for difference. The major world religions all have the resources for doing this, for getting people excited about people who are different from them. It's not every brand that exercises that prerogative, but in the scriptures and traditions of every world religion that has been successful on a grand scale, there is a story about the love of difference.

TIPPETT: Compassion towards difference.

McCULLOUGH: Right. Compassion towards difference. Caring for the strangers in your midst. Being able to see beyond superficial differences toward the essential commonalities.

Religion is also good at appealing to people's meaner sides and the more brutish side. The resources are there for both. So it's really up to those people who have a passion for reconciliation in their own faiths to make sure that the right tones are struck and the others are more muted.

TIPPETT: Something I've been aware of also is that this word "forgiveness" has a really Christian ring in many ears. I remember speaking with a Holocaust survivor who said that, for him, the word "forgiveness" just didn't do it. It has this cultural connotation of forgive and forget. But the Jewish phrase "repair the world" compels him in the same way he feels the word "forgiveness" compels Christians.

MCCULLOUGH: I like that. I wish we could come up with a completely new word for what this human trait is.

TIPPETT: Other than "forgiveness"?

MCCULLOUGH: Or maybe find some new way to talk about it so that we could unload a little bit of the baggage from the past. Because some of the baggage is that it's a namby-pamby thing that doormats do or wimps do. You know, only Milquetoast types of people are interested in it. But from everything I've managed to read and see and understand in my own work, forgiveness is a brawny muscular exercise that I imagine someone with a great passion for life and a great hardy disposition being able to take on.

TIPPETT: And you really feel that it's essential to our geopolitical future, as well as the health of our individual lives.

MCCULLOUGH: It's just too important. It's just too important. And the doors are open now for the use of this kind of language in the public sphere.

8

✳

Knowing How to Heal Ourselves

"STRESS AND THE BALANCE WITHIN"

$\mathcal{E}$sther Sternberg is a scientist's scientist. She has always been wary of the commercialized self-help industry and of unsubstantiated claims for alternative methods of healing. Until she began to do the research we explore here, she shared her profession's modern bias that emotions—such as the gamut of "feelings" that we associate with stress—are distinct and perhaps altogether separate from physical health. Without measurable and logical proof of their direct connection to disease or healing, such a correlation could not be taken seriously.

But in recent years, parallel to her colleagues in many other disciplines, Esther Sternberg underwent a period of scientific and personal discovery. While her mother was dying of cancer, she urged Esther to explore not only whether stress can make us sick, but whether "loving" and "believing" can help us to live well. Sternberg began to pose these questions for herself when she became exhausted and simultaneously developed a form of arthritis, a disease she studies. Here she tells part of her personal story and some of the fascinating history of medicine she traced for the book she ultimately wrote: *The Balance Within: The Science Connecting Health and Emotions.*

Esther Sternberg insists that we'll always need different "languages" to discuss medical fact and emotional realities. And yet for a thousand years "the balance of the four humours"—blood, yellow and black bile, and phlegm—was a central principle of medical teaching. These were visible secretions and therefore could be taken as windows into the workings of the body. Vestiges of these concepts, Sternberg points out, are buried in words we still use to describe emotional types: sanguine, melancholic, phlegmatic, choleric. Modern scientists are now on the cusp of a new world of understanding, she says, because they now know genes, hormones, and neurotransmitters to be as real and measurable as blood and bile. They know that what we call "feelings"—both physical and emotional—are caused by myriad biochemical connections.

This conversation leaves me with a helpful and unexpected appreciation of the positive function of the human stress response. It is as old as time, part of our body's built-in capacity to guide us in new environments and protect us from danger. Stress does not make us sick, per se. But prolonged stress sets off a cascade of reactions that can leave us with overstimulated or suppressed immune systems. Memory and perception add to those physiological effects. Knowing such details, we can understand when we need to seek medical care and when and how we can help to heal ourselves. Such an approach is at the core of the burgeoning field of integrative medicine.

There is a healing paradox in Sternberg's perspective. Science—with its insistence on what can be seen and measured—took us away from our ancient intuition about the connection between health and emotions. But science now is bringing us back. Her insights validate the difficulty of the experience of prolonged stress so many of us know. They illuminate the full meaning of the phrase "feeling sick." She even suggests a notion contrary to our culture of constant productivity: that vacations are not luxuries but physical necessities. So, too, are practices that calm and renew our emotions and our spirits together.

Can stress make us sick? Can places of peace, prayer, meditation, rest, music, and friendship help us to live well? Each of us must answer these questions in the context of our lives, with our particular histories and our physical and spiritual details. But what interesting times we're living in, when physicians and scientists begin to ask such questions along with us.

＊

Stress and the Balance Within

KRISTA TIPPETT, host
ESTHER STERNBERG, immunologist and author

＊

*E*sther Sternberg grew up in Canada, where her father was a professor of medicine. As a child, she knew the Canadian researcher Hans Selye, who coined the medical term "stress" in the 1950s and inserted it into the vocabulary of many world languages.

Today, Sternberg is a leader in the field of neural-immune research, internationally recognized for her discoveries about how the central nervous system and the immune system interact. She's broken new ground in describing how the brain's hormonal stress response might contribute to diseases such as rheumatoid arthritis and depression. In her 2001 book, *The Balance Within*, she explored the history of medicine to understand why, until very recently, modern science failed to treat human emotions, including the feeling we call "stress," seriously. She believes that this drove patients away from some of the sophisticated insights that science can provide. Sternberg first discovered how deeply she herself held that scientific bias when she was asked to write an article on the science of the mind-body connection for *Scientific American* magazine. This process of discovery eventually led to a change in her perspective on science and life. Here is how her book *The Balance Within* begins:

> Nestled at the top of a brown stony hill above the modern Cretan village of Lentas, at the intermingling of cool sage mountain air and warm salt sea breezes, are the ruins of an ancient temple to Asclepius, the Greek god of healing . . . It is a few meters above what was once the source of a natural spring; ancient priests used these waters, and prayer, music, sleep, and dreams to cure the sick . . . And the village people, who still live as one with the rhythms of the sea and sun, know, as their ancestors knew, that emotions and health are one.
>
> As the wind and sun eroded that first ancient shrine, and dried its healing source, something also happened to the world beyond the village. Our faith in the healing power of the spirit also waned; and the god of science and medicine became a much harder, more impersonal god than the fatherly Asclepius. When did we modern scientists and physicians lose the knowledge that was so much a part of these ancient teachings of medicine? And why has the road back to acceptance of this wholeness taken so many centuries to travel?

STERNBERG: Those temples were built at tops of hills overlooking the Aegean or the Mediterranean with beautiful views, always near a freshwater source. The ramps to these temples to Asclepius were built in a long, low slope so that people who could not walk would be helped up these slopes easily. But more important than anything else, there were social interactions; rich social interactions. These were places of healing, like modern-day spas in a way.

But I guess I should go back to the *Scientific American* article, which is how I got to the temples to Asclepius in the first place. I was editing it by my mother's bedside while she was dying of breast cancer. I would be sitting there with my laptop on the armchair in the room. And every time she'd wake up, she'd look at me and ask me what I was writing about and engage me in these very animated discussions. She was a very feisty lady, and she would not let go of the topic. She asked why I just focus on stress and disease: "Why aren't you putting something in there about belief and healing?"

TIPPETT: What did she mean when she used the word "belief"?

STERNBERG: So, that's very interesting. When I was young, she didn't really practice Orthodox Judaism in any way. My grandmother was very Orthodox. But after my grandfather died, and then after my grandmother died, and certainly after my father died, my mother became more and more observant. Going back to the way she'd been raised, lighting the candles every Friday night on the Sabbath, and I think praying in her own way, although it wasn't really overt and open.

The other funny thing is, her nurse in the hospital was a Hasidic Orthodox-practicing Jewish lady. And the two of them would gang up on me while I was writing this article. The Hasidic nurse, of course, knew all the scriptures and came out with all sorts of arguments why I should be including belief and healing in it. My mother was on the emotional side. And I would take the scientific side that this is not proven. This is not something I can put in this article. I was very stern about it. So I published the article my own way, which was talking about stress and illness. Now really in large part, it's because when I was writing this in 1996, this field was still not accepted—not the connection between belief and healing or even stress and illness.

TIPPETT: It's fascinating is that in such a short time, the field has opened up and exploded.

STERNBERG: And I think it's because the scientific research has opened up and exploded. That's really what happened. We have found our way back through the language of science.

TIPPETT: Which also is fascinating—that science forgot or couldn't incorporate emotions and belief into its understanding of what was measurable and real?

STERNBERG: Exactly.

TIPPETT: And it is science that has now established that connection.

STERNBERG: Right. So one of the conclusions that I came to in the book, and also by working on an exhibition at the National Library of Medicine—I learned a lot from that about where this break came, what happened exactly. Scientists need evidence. We need measurable proof. That started with Descartes in the 1600s. At that time, four hundred or five hundred years ago, science didn't have the tools to measure something as ephemeral and abstract as an emotion.

You can measure disease. Disease is an abnormality of anatomy. So when the anatomists of the sixteenth century started to dissect the human body, they discovered that when there was a pneumonia, there was a hole in the lung. When there was a problem in the liver, there was an anatomical anomaly in the liver. The assumption became that disease is associated with an abnormality of anatomy, which allowed huge advances in medicine. In the nineteenth century, Laennec developed the stethoscope so that you could hear problems in the lung. Without seeing the lungs, you could actually hear them. That's concrete; that's easy to understand.

But until very recently, we didn't have the tools to see the living human brain at work with neuroimaging. We didn't have the technology to see how the nerve cells function, the biochemistry, the nerve chemicals that are released, the electrical activity that changes. And only very recently have we been able to see into the genes that make these cells function.

✽

Modern scientists know genes, hormones, and neurotransmitters to be as real and measurable as blood and bile. And the brain as we can see and explore it now, Esther Sternberg says, is not so much one organ as a number of interconnected organs. For example, the human instinct to be alert and vigilant in an unknown environment is controlled by two very different parts of the brain: the hippocampus that controls memory, and the amygdala that controls anxiety and is also known as the fear center. Both of these have connections to the brain's stress center. The complex feeling we know as stress was first named in the mid-twentieth century.

✽

STERNBERG: In the 1940s and '50s, physiology was really reaching its peak. The technologies were available to measure electrical inputs or outputs and physiological responses of the blood vessels and the heart and also hormones. People were— scientists were—beginning to discover hormones. Hans Selye was a physiologist who really borrowed the word "stress" from the physicists, and used it in the biological sense that we know today. He was a very colorful character.

TIPPETT: Didn't you know him?

STERNBERG: Yes. My father and he were professors at the University of Montreal in the Department of Medicine. I put my memories together with talking to his students and colleagues about his theories of stress, which were very revolutionary at the time. His concept was that stress is the body's nonspecific response to any demand. He had mapped out the hypothalamus, the pituitary, the adrenal glands, and he even put in the immune system at that time. He proposed that there were hormones that came out of the hypothalamus, the pituitary gland, and the adrenal glands that would have an effect on how the immune system worked.

People have asked me, So what's different about that? What have we learned in fifty years that Hans Selye didn't say before? Well, we've learned a number of things. First of all, in those days, people who thought about this system stopped at the hypothalamus, which is a very deep structure. It's a structure that's present in all animals. It's a very ancient structure. It's a reflex response, just like your knee jerk. You don't have to think when you're stressed, right? This is a good thing, because if you're driving down the street and a car comes out of nowhere, you don't have time to write a thesis to say, Am I going to put my foot on the brake or not? You have to do this in a millisecond.

TIPPETT: So that's a positive function of stress.

STERNBERG: Right. Now, something very important happens between the bad thing that happens to you, the stressful event, and your physiological response that you recognize as stress. The thing that happens is perception, your perception of that event as stressful. Going back to Hans Selye, the physiologists of the

1950s didn't have the tools to really understand how the rest of the brain was working. So they focused on those deeper parts of the brain, those structures such as the hypothalamus and the adrenaline-like nerves and how they affected the rest of the body. We have advanced to the point where we can really understand much, much better how those inputs, those signals from the outside world, get interpreted by the brain, by all these different parts of the brain. And how they get the overlay of memory on it, so that your memory of certain events can color whether you perceive an event as stressful or threatening or happy.

TIPPETT: And that gets into the life you've lived, the habits you have?

STERNBERG: Everything.

TIPPETT: How healthy you are mentally.

STERNBERG: Right. And that's the part we can hope to change. I want to just say one other thing about Hans Selye. He coined the word, as I said, "stress." And he went around the world getting that word into the dictionary of virtually every country. So that when I was in Japan last year, I asked this audience of mostly Japanese speakers, "How do you say stress in Japanese?" And they said, "Stress." I said, "Well, I guess I speak Japanese." It's in every dictionary. He was very aggressive in doing this. And the sad thing about it was, he also talked to the lay public a lot, and the lay public, of course, loved this. But as a result, his colleagues really disparaged him.

TIPPETT: I spent a lot of the eighties in Germany. And I remembered, as I read your book, the term *der Stress*.

STERNBERG: Yes, yes, *der Stress*.

TIPPET T: But also what's fascinating to me about that is that human beings have experienced what we now call "stress" forever.

STERNBERG: Oh, yes.

TIPPETT: I mean, we know this biochemically. We also just know it in the nature of being human. But we didn't have a word for it in any language?

STERNBERG: Well, it was called different things. In the nineteenth century, it was called "nervousness." Actually, there is a quote of George M. Beard in the 1880s on the principal cause of nervousness in modern civilization—that there are five causes: "the periodical press, the telegraph, the steam railroads, the sciences, and the mental activity of women." So people have perceived things as stressful for a very long time. This is not being facetious; he was describing the stress of the Industrial Revolution. And you could transpose all of those pieces to today.

TIPPETT: Right. Fill in the blanks.

STERNBERG: The media—sorry about that—the Internet, a constant connection with cell phones. And the sciences, because there's all this unknown . . .

TIPPETT: And all of these ethical dilemmas being presented by cutting-edge science that people are facing in a doctor's office.

STERNBERG: Same thing, exactly. And the mental activity of women—I think what he was talking about there is the social change that comes along with technological change, especially rapid technological change. We're living in an information age. Now, why is it these things are stressful? Because change, novelty, is one of the most potent triggers of the human stress response. And that's a good thing when an animal finds itself in a new environment. When a field mouse wanders into a new field, if it didn't have a stress response, if it wouldn't suddenly sit up and look around and become vigilant and focused and ready to fight or flee, if it just went to sleep, it would get eaten by the next cat that came along. Right?

You need your stress response to survive. And novelty must, therefore, trigger the stress response. The problem happens when the stress response goes on too long, when it's active when it shouldn't be active, when you're pumping out these hormones and nerve chemicals at max. That's when you get sick, and that's when these chemicals and hormones have an effect on the immune system and change its ability to fight disease.

Some of Esther Sternberg's most important work has been in determining when stress moves from good to bad as far as the immune system is concerned. If a stressful stimulus or environment is sustained over time, stress hormones and chemicals such as cortisol flood the body. Stress changes us as our minds and bodies, the nervous system, and the immune system communicate and interact. This interaction is too complex, Sternberg believes, to be adequately addressed by the "culture of self-help" that Americans have embraced. At the same time, she says, new science is driving the medical profession to take popular convictions about the mind-body connection far more seriously.

TIPPETT: So what you're saying is, it's not narrowly true—which is the way a lot of us have internalized it—that stress makes you sick.

STERNBERG: Right. It's not the stress that makes you sick. It's that the stress response, those hormones and nerve chemicals, go to the immune system through the bloodstream, through the nerve endings. They hit immune cells that are nearby and change how immune cells work.

TIPPETT: So that same response also dampens your immune system?

STERNBERG: You're pumping out all these hormones. That activates you. That's giving you the stress response. That's making you fight or flee, it's giving you the energy. But that cortisol, that hormone from the adrenal glands, is also the most potent anti-inflammatory drug that our body makes. Now why do I say drug? Because cortisone is the pharmacological form of cortisol. The Nobel Prize was given in 1950 for the discovery that cortisone could be used as an anti-inflammatory drug for arthritis. It didn't occur to scientists and physicians at that time that that wasn't just a drug; that was the body's own way of tuning down the immune response so it didn't go out of control.

TIPPETT: Okay. So something like arthritis is, in fact, an overactive immune system?

STERNBERG: Correct. Arthritis, Crohn's disease, inflammatory bowel disease, lupus—these are all overactive immune responses. You need your immune cells to be active, to create inflammation to fight bacteria, to chew up the bugs, take them away, get rid of them. But then the immune system has to turn off. It has to have an exit strategy; it has to go back to sleep. So there has to be an on-off switch. And there are on-off switches within the immune system. But it turns out that the nervous system plays a very important role in this on-off switch. And there are actually certain nerve chemicals that turn immune cells on, and there are certain ones that turn immune cells off. And cortisol happens to be one that turns immune cells off very powerfully.

The discovery I made is that there can be an actual problem in that circuit that predisposes to developing arthritis. It doesn't mean that stress is causing arthritis. It's that the on-off switch is not working right. It's either stuck in the on position or stuck in the off position. In the case of arthritis, it's stuck in the off position because you can't pump out enough of those hormones to shut off inflammation when you need to shut it off.

TIPPETT: Okay.

STERNBERG: Now, the other side of the coin is when you're chronically stressed—and this is work by the Glasers at Ohio State, Jan Kiecolt-Glaser and Ron Glaser. They've shown that in chronic caregivers of Alzheimer's patients or in people undergoing marital stress, where you're chronically stressed and you're chronically pumping out these stress hormones that are anti-inflammatory, your immune cells are going to be bathed in this anti-inflammatory milieu and will be therefore less able to fight infection.

TIPPETT: Okay.

STERNBERG: So then if you're exposed to a flu bug, a virus or a bacteria, you're less able to fight that. You're more likely to get sick from infections. You're less able to make antibodies when you get vaccines. And it takes twice as long for wounds to heal. So there's no question that when these connections are out of balance, that's when you get sick.

TIPPETT: Does all of this knowledge then kind of reduce us to a mass of chemicals or does it give us more control?

STERNBERG: That's a very good question. For me, it gives me more control. And when I go around speaking to general audiences, I've asked people that. You know, we don't have all the answers now, certainly, to various diseases. But at least we know this is real. Is that good enough? I think people are relieved to know that all these feelings that they've had are real, that we can explain it with nerve chemicals and nerve pathways and hormones and so on. It's not all in your head. You're not crazy if you say that "stress made me sick."

Now, that doesn't mean you shouldn't follow the latest advances in medicine. Of course you should. But understanding these principles, I think, allows you to give yourself permission to take care of yourself. For instance, if you're a caregiver of an Alzheimer's patient, if you understand that by pushing yourself to the max, you're going to really have physiological burnout—not just psychological burnout, that you yourself will get sick—I think it should be easier to then not feel so guilty about giving yourself a break, getting help, getting social support, taking a vacation.

TIPPETT: Or just not saying, "I shouldn't be feeling this way."

STERNBERG: Right. These things are real. And also to know that it's your biology. Another problem with the self-help movement is that I think people feel if they can't fix it on their own, they've failed and they feel bad about themselves. It's important to know that there's a biology to it. And that if you come to the point where you really are in such distress, you do need to seek professional help from somebody who knows how to treat all aspects of stress-related illness.

TIPPETT: You say something interesting. That someone like you— that all of us, really—with the new vocabulary of science, we can talk about emotions and disease, about each of these things as real. And yet that we still need different languages, or that we possess different languages, for describing them.

STERNBERG: When I speak of the language of science, I mean that scientists need evidence. We need hard evidence, data. We have to be able to measure something to know that it's real. So going back to the anatomists, if you couldn't see it, it wasn't real. Actually, until very recently, if you couldn't see it, it wasn't real. But now we have different ways of "seeing" things.

❋

From Esther Sternberg's book *The Balance Within: The Science Connecting Health and Emotions.*

Emotions are always with us, but constantly shifting. They change the way we see the world and the way we see ourselves. Diseases come and go but on a different time scale. And if they change the way we see the world, they do it through emotions. Could something as vague and fleeting as an emotion actually affect something as tangible as a disease? Can depression cause arthritis? Can laughing and a positive attitude ameliorate, even help to cure, disease? We all suspect that the answers to these questions are yes, yet we can't say why and certainly not how. Indeed, entire self-cure industries have been built on this underlying assumption. But physicians and scientists until recently dismissed such ideas as nonsense because there did not appear to be a plausible biological mechanism to explain the link.

. . . Part of the reason for this is that scientists and lay people speak different languages—but so do emotions and disease. Poetry and song are the language of emotions; scientific precision, logic, and deductive reasoning are the language of disease.

✳

STERNBERG: So maybe I should go back to my own experience in this. I was very focused on these very endocrinological, molecular neuroscience studies and analyzing all the different nerve pathways that are involved in the stress response, and the differences in arthritis-prone rats and arthritis-resistant rats. But the converse of that, the corollary of that, is that if you understand how breaking the connections can make you ill, then perhaps you can figure out how you can fix those connections. Right? If you're pumping out too much of those stress hormones, so you're chronically stressed and your immune system is tuned down, what are the things that you can do intuitively to reduce that, to get that back to balance? And these are the things that are being worked on now and are certainly being worked on increasingly. You can begin by thinking about taking yourself "off-line," so you're not chronically stressed.

TIPPETT: What do you mean by that analogy?

STERNBERG: When your computer is jammed up, when your e-mail is jammed up with spam, what do you do? You shut down and reboot. Right? We know this about computers. We don't seem to know it about our bodies and ourselves. Shutting down and rebooting is a really important thing to do. So if you're working 24/7 on a deadline and you're exhausted, if you're a chronic caregiver of an Alzheimer's patient, this means taking yourself away from that situation as much as you can.

In my own case, when I was writing that article by my mother's bedside, and in the last throes of her illness flying up to Montreal all the time and under a huge amount of stress, and then after she died going through the grieving process—around that period, I became sick myself. I developed an inflammatory arthritis. Of course there were the genes in the family; these diseases don't just come from stress. There has to be some predisposition. But the question is, Why did I develop it at that moment? Why didn't I develop it five years before or five years later? I believe there's no question, and there's evidence to support the notion, that being chronically stressed can be associated with triggering these sorts of diseases from burnout.

TIPPETT: Did it help you, as that happened to you, because you knew something of the biochemistry?

STERNBERG: I guess what happens when you understand the anatomy and physiology of the system is that you can stand back and become an observer of your own situation. You can, to a certain extent, treat yourself as the patient and dissociate yourself from yourself. But then there's the patient's side of you that really doesn't feel great.

TIPPETT: It doesn't go away.

STERNBERG: It doesn't go away. And you may not do the right things. So, even knowing all this stuff, I didn't stop. I just burned myself out, effectively.

✸

In the midst of exhaustion and illness, compounded by a move, Esther Sternberg received a surprise invitation from new neighbors to stay with them on the Greek island of Crete. It was through that invitation that Sternberg found herself at the site of the ancient temple of the Greek god of healing, Asclepius. He and his daughters, Hygieia and Panacea, symbolize the timeworn human insight that health lies in balance. This experience changed and framed the conclusions Sternberg ultimately drew, both medically and personally.

✸

STERNBERG: The more I thought about it, the more I thought that there are a lot of indirect pieces of evidence that one can piece together to construct a logical argument that believing makes a difference. We're not talking about what you're believing in or whom you're believing in—but the act of feeling spiritual, maybe that's what I really mean, that feeling of wonder and awe that one gets when one is in a spiritual place, that thrill of seeing a sunset. When my sister and I were small, we lived at the base of Mount Royal, which is a hill in Montreal. And whenever it looked like there was a beautiful sunset, we would drop everything—we were washing the dishes, we were having dinner, whatever—we'd pile into the car, drive up to the University of Montreal, which was on the top of one of the hills, and look at the sunset.

I assumed that everybody used to do this as a child. It inspired in us a sort of awe at nature and life and beauty. My parents had been through the war. They had both been born in Romania, and my father had been in a kind of starvation, concentration, camp in Russia. He was a physician; he managed to get out. And my mother had gotten out before the war. But it was very palpable that life and peace meant a lot to them. I remember sitting on summer mornings on the terrace at home with my father, early in the morning having breakfast, and he would look up—he used to read a lot—and he'd look up and he'd say, "Stop, listen. Listen to the sound of peace." You'd hear the birds chirping and the dog barking and the tennis balls on the tennis court across the way. And all those things, I guess, became part of me—to understand that you can find a place of peace if you stop and look and listen. And I think that's what happened to me in Crete. So when I went with my neighbors to Crete, to their little cottage. I did bring my laptop, but unfortunately, the . . .

TIPPETT: Voltage?

STERNBERG: Yes, voltage—you got it. I had the converter, whatever, for the voltage, and it blew on the first paragraph. They never forgave me. They said, "I thought you were going to be writing your book." But I wrote it in my mind. And I enjoyed the place. I would swim every day in these warm, wonderful waters of the Mediterranean. At first I couldn't walk very well. But by the end of the time, I was able to climb up these hills of sort of scrabbly rock. And then I would climb up to the top of the hill above the town, which had these ruins of the temple to Asclepius, the Greek god of healing. And on top of the temple site, as is pretty typical, there was a Byzantine church. Then on top of that, there was a tiny little Greek chapel that was modern, but I think it was about three hundred years old. There were all these icons and candles to the icons. It was just the most amazing peaceful place. And I would sit out there and look at the ocean and stay for hours, crawl around the ruins and look at these amazing things. It gave me a sense of peace, really a sense of spirituality of place and time. And when I came back to Washington, I didn't need to go into the hospital.

Now, you could argue, and the physician side of me says, well, I had been put on high-power medication before I left.

TIPPETT: So it may have kicked in?

STERNBERG: It took some time to kick in. But I asked before, Why did I get sick at that very moment when I was stressed? So why did I get better in such a relatively short time after this period of rest and social support, healthy diet, and beginning gradually to exercise more and more? Why did I get better then and not a month later? I think I was allowing those medications to kick in, because I wasn't forcing my body to work against them.

TIPPETT: Right. You don't see the efforts that we can make to manage stress as an alternative to medicine but as a partner to medicine.

STERNBERG: Oh, a partner, absolutely.

TIPPETT: You know what this makes me think, though? Even when you tell the story about your parents, and them helping you appreciate a peaceful moment or a beautiful sunset, we don't all have that. And even though there are ways to use this knowledge we have about what's happening in our bodies, and how that's connected with our emotions, we come to this knowledge with different wounds and weaknesses and different degrees of damage from our families. You talked about how memory plays a role in this, and how we've been traumatized by memories differently. It seems like a sort of built-in inequity in terms of how we can use the knowledge.

STERNBERG: Well, I think that the memories can go both ways. You can have positive memories that trigger positive emotional responses and a flood of positive nerve chemicals, endorphins, you know, those dopamine reward chemicals. And you can have negative memories that trigger the stress response. A week's vacation isn't going to do it for everybody. It depends on how deep the wounds are, at what stage you are in the grieving process, your genetic makeup, whether you have the genes that predispose to depression or not, whether these kinds of wounds then trigger a biological depression that just can't be fixed with a vacation. This sort of thing needs to be fixed by fixing the imbalance and the nerve chemicals with antidepressant drugs, together with psychotherapy and cognitive behavior therapy, working on those memories. That's what psychotherapy is about, digging deep into those memories. You can't do it overnight.

TIPPETT: We haven't talked about psychotherapy. How can psychotherapy fit into this whole picture of what's going on with us when we're not functioning as well as we could or as healthy as we could be?

STERNBERG: My way of thinking about it—and now we're getting more into speculation than into science—but when you think about meditation, that's another thing that changes the way the brain works.

TIPPETT: And we can measure that and see that.

STERNBERG: We can measure that, right. Richie Davidson at the University of Wisconsin has done this with meditating monks, those sort of Olympic meditators, and there are different parts of the brain that become active and different parts that shut down. Meditation is a state, just like being awake or being asleep, but it's a different state than being awake or being asleep. We don't fully understand exactly what happens when one is meditating. But clearly, there are different nerve chemicals released in those states. And there's evidence to believe now that meditation can change how your immune system works, probably through these nerve chemicals. So meditation is one, psychotherapy is another, yoga, exercise, when you have a runner's high. I swim, and after about ten or fifteen minutes of swimming, you get into this peaceful zone. I think what's happening in all of these settings is, you're relearning how to perceive that stressful event. If you think of learning how to ride a bicycle, the first time you get on a bicycle when you're a kid, you fall off. Right? You need to get on about fifty times. Okay? It's known that if you're going to learn something, you have to do it repetitively about fifty times. That's why your mother told you to practice the piano every day.

TIPPETT: Okay, okay. And you're saying the same thing that happens in your brain when you practice piano that you finally get it, it's like what can happen to . . .

STERNBERG: You finally get it. There's an "aha" moment, yes.

TIPPETT: . . . in your perception, your response to stress.

STERNBERG: In psychotherapy, you can go over and over and over those same loops, and your therapist can tell you you're doing that—which many of them don't because they're trained not to tell you consciously what's going on there—but you need to come to it yourself after going over and over and over it about fifty or maybe more times. Then you suddenly get it.

TIPPETT: Looking around the world we live in, the culture, American culture—having named the word "stress" or invented it, we now probably overuse it. Everyone I know feels overwhelmed by stress. So do I. What would you wish for us? What are some simple things that you would like to see happening that you think could make life feel more manageable?

STERNBERG: I think you're right. We do live in an era that is filled with very rapid technological change, and so in that way is like the Industrial Revolution that I described before. But we also live in a fearful world, which for Americans is a relatively new thing. It always used to be over there. Since 2001, September 11, it's come here. The rest of the world has lived with this for many, many centuries. We've had the privilege to not have to deal with the fear of the unknown day to day. But there's no question that there is a lot of fear and stress out there. And it may not always be possible, but I do think we need, each one of us, to find our place of peace and try to go there every day. We take our cars in to be serviced every five thousand miles, but we don't do that with ourselves. I'm sure it's different for every person. Some people may find it through meditation, some through prayer, some through yoga, some through exercise, some through music, some through reading, some through art, whatever it is that does it for you. Any amount of time that you can devote to going offline, in whatever way you find, will help.

9

*

The Nature of Human Vitality

"THE SOUL IN DEPRESSION"

W e're increasingly aware in our culture of the many faces
of depression, and we've become conversant in the language of
psychological analysis and medical treatment for it. But depression has a profound spiritual effect that is much harder to speak
about and can often be traced only years onwards. Still, like many
things that are difficult to speak of—and given the epidemic scale
of this illness in modern Western society—this is important reflection for our common life. I feel that the three voices in this
discussion bring rare, brave, and helpful insights into the light.

I took these conversations as an occasion to walk myself with
some trepidation back through the spiritual territory of despair.
Like many millions of people, I have experienced severe, clinical
depression. And I think that "depression" is one of the most misleading and inadequate words in our vocabulary. When I try to
describe the experience, I find myself grasping to say what it is
not. Depression is not essentially about being sad, or down, or
blue, though these may be symptoms. In the illuminating language of Andrew Solomon, the opposite of depression is not

happiness—it is "human vitality." It can have purely physiological origins. It may be triggered by old sadnesses grown unbearable or anger turned inwards, as one saying goes. But it becomes a way of being in, and moving through, the world.

Ignatius Loyola, the sixteenth-century founder of the Jesuit order, spoke of "desolations"—a better word than depression, in my mind—that "lead one toward lack of faith and leave one without hope and without love. One is completely listless, tepid, and unhappy, and feels separated from our Creator and Lord." For me, depression was not so much about being without faith or hope or love; it was, rather, not being able to remember knowing those things, not being able to imagine ever experiencing them again.

After depression there is a particular solace in the voices of others who've been marked by this disease and lived to reflect on its contours. When I finally began to emerge, I found a kind of comfort in the scriptural psalms of lament and imprecation— mourning prayers, cursing prayers. Suddenly the "pit" of which the psalmist so often writes was real to me. I also returned to the poetry of Rilke, who like many great thinkers and creators, had an intimacy with "darkness." Anita Barrows has luminously translated some of his poetry. Personally and in her work as a psychologist, she has also grappled with depression all of her life. She traces a difficult but ultimately hopeful line between the illness of depression and the darkness that is a part of human vitality and that we can embrace.

But a cautionary word is necessary. In the midst of depression, very little if anything is possible in the way of spiritual reflection. The insights found here are all hard-won and came much, much later after a period of recovery and healing. If you know someone who is depressed now, or if you yourself are in that state, go gently, seek help, and don't expect spiritual breakthroughs. In a section of this discussion that many people describe as helpful, Parker Palmer tells of the friend who helped him most during his worst episode. The friend came and sat silently with him, day after day, and "merely" massaged his feet. In such simple human gestures we find the most essential comfort.

Parker Palmer's two bouts of paralyzing depression came, in fact, while he was a leader of a spiritual community. In the end, that experience reframed his whole understanding of spiritual life. And I'll leave the last words of this reflection to him—on the important question of the role and presence of God in the suffering of depression:

> I do not believe that the God who gave me life wants me to live a living death. I believe that the God who gave me life wants me to live life fully and well. Now, is that going to take me to places where I suffer because I am standing for something or I am committed to something or I am passionate about something that gets resisted and rejected by the society? Absolutely. But anyone who's ever suffered that way knows that it's a life-giving way to suffer—that if it's your truth, you can't not do it. And that knowledge carries you through. But there's another kind of suffering that is simply and purely death. It's death in life. And that is a darkness to be worked through to find the life on the other side.

*

The Soul in Depression

KRISTA TIPPETT, host
ANDREW SOLOMON, novelist
PARKER PALMER, Quaker author and educator
ANITA BARROWS, psychologist and poet

*

As a society, we're increasingly aware of the many faces of depression, and we've become conversant in psychological analysis of depression and medical treatment for it. But there is a growing body of literature by people who've struggled with depression and found it to be a lesson in the nature of the human soul. Such insights are scarcely possible while one is in the throes of depression, but they can come later after a process of recovery and healing.

I have experienced severe depression. I took the making of this program as an occasion to walk with some trepidation back through the spiritual territory of despair. The voices in this discussion span a range of varieties of depression and religious perspective. Anita Barrows is a poet and psychologist. Parker Palmer is a Quaker author and educator. Andrew Solomon is the author of *The Noonday Demon: An Atlas of Depression*, for which he received the National Book Award and a Pulitzer Prize nomination.

In 1998, Andrew Solomon published an article in *The New Yorker* magazine about his experience of clinical depression. His story elicited over one thousand letters from *New Yorker* readers. In excruciating detail, he described his breakdowns and his extreme immersion in the brave new world of antidepressant pharmacology.

After that article and his subsequent book, Solomon was interviewed widely. What struck me as I listened was how his questioners tended to focus on his physical collapse and not on his eloquent insistence, between the lines, that depression for him was also a spiritually revealing experience. And Andrew Solomon is not a religious person. His mother's death when he was twenty-seven triggered his first major depression. As he recounts in *The Noonday Demon*, she committed a planned suicide in the presence of him, his father, and his brother to end a bitter struggle with cancer.

He traces the onset of his depression from his incapacity to grieve the death of his mother.

SOLOMON: The passage from grief into nothingness was very alarming and very strange. I still would have said, I'm terribly upset that my mother died, and so on and so forth. But the feeling went out of it. I think that's why, when the feeling comes back, you think: this is a soul. This is a spirit. This is something profound and alive which returned to me after taking a leave of absence.

TIPPETT: What I found really refreshing about your book—and something I don't think is out there enough—is what depression really is and what it really is not. It's not sadness. I think you say that the opposite of depression is human vitality.

SOLOMON: It's an experience, overall, of finding the most ordinary parts of life incredibly difficult: finding it difficult to eat, finding it difficult to get out of bed, finding it difficult and painful to go outside, being afraid all the time and being overwhelmed all the time. Frequently, it's quite a sad experience to be afraid and overwhelmed all the time. Nonetheless, those are the essential qualities of it. It isn't primarily an experience of sadness.

TIPPETT: Right.

SOLOMON: And it teaches you how big emotion is. The profundity of the inner self, I suppose, would be the best way of putting it.

TIPPETT: Are "passions," in a classical sense of that word, also a way to talk about the largeness of emotion that you're describing?

SOLOMON: I think passions are the only way to talk about it, the passion which is the essential motivator for all human activity. In a sense, after you've been through a depression, it gives you a different relationship to the world. It gives you a different sense of how your interior monologue really determines everything. And you're left mystified as to where that interior monologue originates and where those passions come from and why they're so mutable and what it is within them that's immutable.

TIPPETT: I'd like to talk about medication. You are still on medication, I believe and, I suppose, will be forever, which is becoming an advisable way for people who've suffered multiple depressions.

SOLOMON: That is right, yes.

TIPPETT: What kind of regimen of medication do you live with now?

SOLOMON: Well, I'm in the process of shifting things around, because at the moment I'm on really more than I'd like to be. But right now I'm taking Lamictal, Zyprexa, Lexapro, BuSpar, and Wellbutrin.

TIPPETT: I wonder if people ask you, How do you know that this person you are now and even this wisdom that you have, that this is really you, when you are so influenced by chemicals?

SOLOMON: I think the idea that there is a real self and that changing it in any way with medication is artificial is like the idea that you really have teeth that fall out when you're thirty and that you're artificially changing them by using modern dental care. I just think the authentic thing goes through periods of flaw and illness and problem, and that you have to address those problems. Taking these medications brings about effects, which are also brought about by certain kinds of talking therapies and external experiences. And I'm a great believer in those therapies and also continue to work in those areas and arenas.

There's a lovely passage from *The Winter's Tale*, which I quote toward the end of the book, beautifully phrased, and I wish I had it in front of me. I'd read it out loud.

TIPPETT: Here's a sentence I think may have been from that passage or your commentary on it: "If humanity is of nature, then so are our inventions."

SOLOMON: Yes, exactly. And it ends, that passage, with the line, "The art itself is nature."

TIPPETT: You also quote the poet Jane Kenyon: "We try a new drug, a new combination of drugs, and suddenly I fall into my life again." From my own experience, I remember that. And again, it

is so hard for people who haven't been through this to imagine—
that it is not like you are changed into someone new, but you fall
into your own life again. So mysterious.

SOLOMON: I feel that very strongly. I think I relate this anecdote
in the book. There's somebody I used to know, and I ran into
her in the street. I said, "How are you doing?" and she said, "Well,
I had a very serious depression." And I said, "Are you taking
medications? Have you been in therapy?" She said, "No, I just
decided it was the result of stress. So I eliminated the stresses
from my life. I broke up with my boyfriend because that was difficult. I gave up my apartment to live in a one-room place, because I thought that would be less demanding. And I don't really
go out to parties anymore, because I find being with people is
just very difficult for me." She went on and on with this catalog,
and I thought, "That is not true to yourself. I've known you for
years, and you are a different person."

I feel as though I've made, in effect, the opposite decision. I
have the personality that is consistent with the personality I had
when I was ten and twenty and twenty-five, and that then began
to fall apart a little bit later on. And I have a strong sense that the
medications have returned me to myself.

Andrew Solomon's award-winning book *The Noonday Demon* is
at once a memoir and a compendium of the many nuances of
depression, described from medical, scientific, and social perspectives. He also delves into historical attitudes towards depression, including religious ideas, which have formed modern
attitudes in the West.

Many ancient classical thinkers did not detach the psyche
from the body. By contrast, the great fifth-century church father

St. Augustine labeled depression a disease not of the body but of the soul, and a mark of God's disfavor. This Christian stigma, Andrew Solomon says, has remained in modern America even when the theology behind it has not. I asked him what, if any, religious literature he has found to be helpful.

✳

SOLOMON: I think I would say that I found a particular comfort in the harder rhetoric of Judaism, though I vastly appreciate the more forgiving nature of the New Testament. But the Old Testament had a certain doctrine of acceptance and law and endurance that these terrible things happen, and you just stick it out, and maybe they get better and maybe they don't get better. One would expect in a depression that what one needs is softness, and I think one does need softness from other people. But I found those basic lessons, which I had absorbed in those Sunday school lessons when I was a child, had a sternness in them that I found very believable even when I was at my lowest. At a time when I couldn't have believed that God loved me, I could believe that there was logic and structure in the world. And so for me, as a Jew, I think that was a particularly potent comfort and guide through what was happening.

TIPPETT: That's fascinating because on the surface, it doesn't sound—You would think that those passages especially might alienate a modern person, a sophisticated, educated city dweller.

SOLOMON: They're much easier to believe if you're a sophisticated city dweller.

TIPPETT: You write, "Depression is the flaw in love." What do you mean by that? It's a haunting sentence.

SOLOMON: It seems to me that, in a way, the most fundamental and important capacity we have as human beings is the capacity for love. And the feeling of love couldn't exist without a range of other feelings that surround it, the primary one being the fear of loss. If the loss of someone you love didn't make you sad, then what substance would the love have? Therefore the emotional range that includes great sadness and great pain is essential to the kind of love and attachment that we form. It seems to me that the kind of severe depression that we've been talking about represents an overactivity of the mood spectrum. But that without the basic mood spectrum of which depression is the extreme end, we couldn't have the experience of intimacy, which that brings.

TIPPETT: You also have spoken a lot about how for you the experience of depression and also a recovery of the capacity or a deepening of your capacity for intimacy go together. Does that flow from that same thought?

SOLOMON: Yes, I think it does. I think the awareness of my own vulnerability has made me more aware of other people's vulnerability, and more appreciative of people who cushion me from the things to which I am vulnerable. So I think it's made me both more loving and more receptive to love, and given me a clearer sense than I would otherwise have had of the value of love. And I suppose, again, without wanting to get into a suggestion of specific doctrine, that that has also given me a sense that some abstract love in the world, which I suppose we could call the love of God, is essential and significant. And it has been increased in me, both in terms of my appreciation for it and my feeling of being loved or held.

I use that word "soul" very advisedly. I don't particularly mean something that will eventually acquire wings and go off to the kingdom of heaven. I guess, though, if you say "the mind" or you say all of those things that get used in scientific discussions of depression, like "emotional infrastructure" and other phrases like that, they just seem to me not to capture this essential self.

TIPPETT: Those are too clinical.

SOLOMON: And it seems to me that who other people are is always mysterious. What I realized in the wake of depression is that who I am is fully mysterious to me. And since I don't fully know it and since I can't fully comprehend it—it's not simply that I don't, it's that I can't—then there has to be some mystical element in it and some element that's obviously present and yet beyond my comprehension. And that is what I was trying to characterize when I used the word "soul." The recognition of that fundamental reality has been much stronger in religious writing and in religious contemplation than it has been in other areas of considering and enterprise.

TIPPETT: You used the word "soul" near the very beginning of your book and right at the end again, I noticed. I'm not sure you used it many other times throughout.

SOLOMON: Yes. That was quite deliberate actually. Given that I didn't want to write a religious book because I am not in any very mainstream way a religious person, I didn't want to adopt the word all the way through. But I felt that it was an important mode of description, and I wanted it to frame all of what I was saying.

✳

I experienced my own severe bout of clinical depression in 1995. My symptoms were classic: sleeplessness, weight loss, fear, anxiety, and a devastating inability to concentrate. In depression, I found body, mind, and spirit to be shockingly, maddeningly inseparable. As I was gradually emerging, I read an essay by the author Parker Palmer, which echoed this experience of my own. But the article surprised me. I knew of Parker Palmer as a guru of the soul, a wise Quaker thinker whose books and speeches had helped many people integrate their deepest spiritual values into their lives and work. And yet here was a revelation by Parker Palmer that he had suffered two episodes of crippling depression in his forties.

When Parker Palmer experienced his depressions, he was the revered leader of a Quaker spiritual community. At first, because of this, he felt ashamed. But ultimately, he says, depression forced him to reconsider the core of his understanding of spiritual life itself.

PALMER: Going into my experience of depression, I thought of the spiritual life as sort of climbing a mountain until you got to this high, elevated point where you could touch the hand of God or see a vision of wholeness and beauty. The spiritual life at that time had nothing to do, as far as I was concerned, with going into the valley of the shadow of death. Even though that phrase is right there at the heart of my own spiritual tradition, that wasn't what it was about for me. So on one level, you think, "This is the least spiritual thing I've ever done." And the soul is absent, God is absent, faith is absent. All of the faculties that I depended on before I went into depression were now utterly useless.

236 ❋ EINSTEIN'S GOD

And yet, as I worked my way through that darkness, I sometimes became aware that way back there in the woods somewhere
was this sort of primitive piece of animal life. Some kind of existential reality, some kind of core of being, of my own being—I
don't know, maybe of the life force generally—that was somehow holding out the hope of life to me. And so I now see the soul
as that wild creature way back there in the woods that knows
how to survive in very hard places, knows how to survive in
places where the intellect doesn't, where the feelings don't, and
where the will cannot.

TIPPETT: Where is God in all of this?

PALMER: Well, Tillich, you know, described God as the ground
of being. I no longer think of God as up there somewhere. I think
of God as down here. In my own Christian tradition, that is
pretty consistent with incarnational theology, with the whole
notion of a God who journeyed to Earth to be among us compas-
sionately, to suffer with us, to share the journey.

TIPPETT: I love this sentence from your book *Let Your Life Speak:*
"I had embraced a form of Christian faith devoted less to the
experience of God than to abstractions about God, a fact that
now baffles me: how did so many disembodied concepts emerge
from a tradition whose central commitment is to 'the Word become flesh'?"

PALMER: That's a baffling question to me to this day. But I take
embodiment very seriously. Depression is a full-body experience and a full-body immersion in the darkness. And it is an
invitation—at least my kind of depression is an invitation—to
take our embodied selves a lot more seriously than we tend to do
when we're in the up-up-and-away mode.

TIPPETT: Let's dwell with that for a moment. There is a critique that Christian tradition does not help people who are suffering from something like depression—that suffering itself, by some interpretation, would be said to be glorified. But you're turning that image around in terms of the way you've come to apply it.

PALMER: Yeah, I am. There's a lot, unfortunately, about suffering in Christian tradition that's hogwash, if I can use a technical theological term. It's awfully important to distinguish in life, I think, between true crosses and false crosses. In my growing up as a Christian, I didn't get much help with that. A cross was a cross was a cross, and if you were suffering, it was supposed to be somehow good. But there are false forms of suffering that get imposed upon us, sometimes from without, from injustice and external cruelty, and sometimes from within, that really need to be resisted.

I do not believe that the God who gave me life wants me to live a living death. I believe that the God who gave me life wants me to live life fully and well. Now, is that going to take me to places where I suffer because I am standing for something or I am committed to something or I am passionate about something that gets resisted and rejected by the society? Absolutely. But anyone who's ever suffered that way knows that it's a life-giving way to suffer. If it's your truth, you can't not do it, and that knowledge carries you through. But there's another kind of suffering that is simply and purely death. It's death in life, and that is a darkness to be worked through to find the life on the other side.

✳

Parker Palmer experienced two crippling bouts of depression in his forties. He recalls a particular thought offered by his psychologist that he says eventually helped him reclaim his life. The therapist said, "Parker, you seem to look upon depression as the hand of an enemy trying to crush you. Do you think you could see it instead as the hand of a friend pressing you down onto ground on which it is safe to stand?" Today Parker Palmer writes theologically about depression. He even traces his own collapse back to his midlife conversion to contemplative Quaker tradition.

PALMER: I went to a friend at one point. She happens to be a member of a religious community, a sister. I said, "I've been on this wonderful Quaker journey. I've been sitting in silence and I've learned to pray, and I've been feeling so much closer to God than I ever did when I was just clinging to doctrine. Why am I now feeling so full of death?" And she said, "Well, I think the answer is simple. The closer you get to the light, the closer you also get to the darkness." That was another one of those phrases, like the one my therapist gave me, that I didn't understand right away. But right away I knew there was some kind of truth in it that I needed to try to understand.

TIPPETT: How do you understand that phrase now?

PALMER: I understand that to move close to God is to move close to everything that human beings have ever experienced. And that, of course, includes a lot of suffering, as well as a lot of joy.

TIPPETT: And just getting back to the subject of this show, the thing in the midst of a depression that feels so absent, I would say, is your very soul, right? The ground of your being has dropped out.

PALMER: Right.

TIPPETT: I don't even think I could think about God one way or the other. I had to put the idea of God to one side. And yet some of the most profound observations that you're making and that you're saying that can be possible out of some depression are precisely about those aspects of human experience.

PALMER: Right. And as I said earlier, as best I can reconstruct it—and a lot of it's hard to reconstruct because I was so out of it that I don't entirely trust my capacity to reconstruct it—but as best I can, the thought of God, all of those theological convictions, were just dead and gone during that time. But from time to time, back in the woods, that primitive wildness was there. And if that's all God is, I'll settle for it. I'll settle for it easily and thankfully.

TIPPETT: When you were talking about how in Quaker tradition people know how to be silent, I was recalling that passage in what you've written about your depression, about the friend who helped you the most, who would just come be with you.

PALMER: I had folks coming to me, of course, who wanted to be helpful, and, sadly, many of them weren't. These were the people who would say, "Gosh, Parker, why are you sitting in here being depressed? It's a beautiful day outside. Go, you know, feel the sunshine and smell the flowers." And that, of course, leaves a depressed person even more depressed. Because while you know intellectually that it's sunny out and that the flowers are lovely and fragrant, you can't really feel any of that in your body, which is dead in a sensory way. And so you're left more depressed by this "good advice" to get out and enjoy the day. Other people would come and say something along the lines of, "Gosh, Parker, why are you depressed? You're such a good person. You've helped so many people."

TIPPETT: "You're so successful."

PALMER: "You're so successful, and you've written so well." And that would leave me feeling more depressed because I would feel, "I've just defrauded another person who, if they really knew what a schmuck I was, would cast me into the darkness where I already am."

But there was this one friend who came to me, after asking permission to do so, every afternoon about four o'clock. He sat me down in a chair in the living room, took off my shoes and socks, and massaged my feet. He hardly ever said anything. He was a Quaker elder. And yet out of his intuitive sense, he from time to time would say a very brief word like, "I can feel your struggle today," or farther down the road, "I feel that you're a little stronger at this moment, and I'm glad for that." But beyond that, he would say hardly anything. He would give no advice. He would simply report from time to time what he was intuiting about my condition. Somehow he found the one place in my body, namely the soles of my feet, where I could experience some sort of connection to another human being. And the act of massaging, in a way that I really don't have words for, kept me connected with the human race.

What he mainly did for me, of course, was be willing to be present to me in my suffering. He just hung in with me in this very quiet, very simple, very tactile way. And I've never really been able to find the words to fully express my gratitude for that, but I know it made a huge difference. It became for me a metaphor of the kind of community we need to extend to people who are suffering in this way, which is a community that is neither invasive of the mystery nor evasive of the suffering but is willing to hold people in a space, a sacred space of relationship, where this person who is on the dark side of the moon can get a little confidence that they can come around to the other side.

✳

Depression runs through the literature and poetry of every culture. In older works, it is often referred to as melancholia. The psalmist of the Hebrew Bible wrote repeatedly of the "pit" of despair. The sixteenth-century Spanish mystic John of the Cross penned the phrase "the dark night of the soul." And there is a growing Buddhist literature on such themes. The Zen teacher and Jungian psychotherapist John Tarrant has written about "the light inside the dark," defining the soul as that part of us which touches and is touched by the world.

Anita Barrows has been a practitioner of Theravada Buddhism for most of her adult life. As a psychologist, she says that the Buddhist embrace of inner darkness can be terrifying and even dangerous in the depths of clinical depression. But like Andrew Solomon, she honors darkness as an aspect of life. Barrows has lived with depression as far back as she can remember—first of all, vicariously, through life with her mother.

✳

BARROWS: My mother would say things like, "I talk to God. I talk directly to God, and he answers me." And I always had the image when I was a child that God was this old man, half shaven, in a bathrobe, who had a direct phone line to Sylvia, my mother, but didn't do very much to help her. I always thought, "If she has such a direct line, why doesn't he make her better?"

The reason I was told for my mother being in bed so much was that she had warts on her feet. It was kind of an odd thing to have been taught. And the warts had a wonderful name, an Italian name. It was *verruca*, which to me sounded kind of like a Hebrew prayer, *Baruch atah*. And so I was fascinated with the word. But I would sit outside the door to my mother's bedroom, and I would hear her crying or just wait for her to wake up. That was very much the experience of my childhood.

I even remember a very strong sensation walking through the door. We lived in an apartment during that middle part of my childhood, from the time I was about seven until I was ten. I remember walking through the door and really feeling a change in the atmosphere from the vivid outside world where I loved to be. Whatever the weather, I loved to be outdoors. And I would walk inside and I would feel a kind of permeable darkness. That was my mother's depression.

TIPPETT: That's an amazing image. You're already getting at something that I want to try to bring into the light, which is that depression is something many of us have experienced either ourselves or through others. And we talk about it from a medical standpoint and from a psychological standpoint. But "permeable darkness" is really a good description of the wholeness of that.

BARROWS: Yes. It was permeable in that I could walk in and out of it myself, and put my hand in it and feel what it felt like. That was certainly something my mother lived with all her life, and it's a state that's familiar to me as well, although I have lived it differently from the way my mother did.

✳

Anita Barrows experienced an early bout of depression at seventeen, after she left home for college. Then after the birth of her first much-wanted child when she was thirty-one, she suffered a major collapse. That depression had an organic cause, an autoimmune disease of the thyroid. After many false diagnoses, it was easily treatable. But like all of us who've been touched by depression, whatever its form, Anita Barrows remains marked by the presence of this illness in her life. And more than most of us, I think, she embraces it actively. She has explored the spiritual aspects of darkness and light through writing poetry and translating the work of others. Together with the Buddhist scholar Joanna Macy, Barrows created a stunning translation of Rainer Maria Rilke's *Book of Hours*. And as a psychologist who is also a lover of language, she complains that the word "depression" itself does not do justice to this aspect of human experience.

✳

BARROWS: It almost becomes a way of dismissing it. I see it much, much more as a kind of a minor-key chord that is a constant accompaniment to one's life.

TIPPETT: To any life?

BARROWS: To many lives. Well, to the life of a person who is inclined in that direction. Rilke loved the darkness, and there are many poems where he speaks about darkness in a way that really, I think, is what drew me to these poems.

"I love the dark hours of my being," he wrote. I mean, there have been times certainly in my life when the depressed mood— it's such a terrible word. The dark mood. It's a word that has taken on so many rotten connotations, you know. It's a medical term now. I want to redeem it from the medical and the clinical. There is a point in depression that is so devastating that only in retrospect would anyone want to say, "I am glad I touched bottom because now I know what that is." But this other kind of living with darkness, which is so familiar to me, I think is a very sort of spiritual place. There is a kind of ripening that goes on in that place, a quieting, a listening, a place of nonactivity.

TIPPETT: And also a loss of illusions about what activity will get you.

BARROWS: Exactly. All you can do in that place is sit and listen and be, and be very simple. Rilke again says, "Be modest now, like a thing ripened until it is real, so that he who made you can find you when he reaches for you."

Here is the poem with that line in Rainer Maria Rilke's *Book of Hours*, which Anita Barrows translated with Joanna Macy and subtitled "Love Poems to God."

> You are not surprised at the force of the storm—
> you have seen it growing.
> The trees flee. Their flight
> sets the boulevards streaming. And you know:
> he whom they flee is the one
> you move toward. All your senses
> sing him, as you stand at the window.
> The weeks stood still in summer.
> The trees' blood rose. Now you feel
> it wants to sink back
> into the source of everything. You thought
> you could trust that power
> when you plucked the fruit;
> now it becomes a riddle again,

and you again a stranger.
Summer was like your house: you knew
where each thing stood.
Now you must go out into your heart
as onto a vast plain. Now
the immense loneliness begins.
The days go numb, the wind
sucks the world from your senses like withered leaves.
Through the empty branches the sky remains.
It is what you have.
Be earth now, and evensong.
Be the ground lying under that sky.
Be modest now, like a thing
ripened until it is real,
so that he who began it all
can feel you when he reaches for you.

❋

BARROWS: Suddenly, in depression you are ripped from what felt like your life, from what felt right and familiar and balanced and ordinary and ordered. You're thrown into this place where you're ravaged, where the wind rips the leaves from the trees, there you are. Very, very much the soul in depression.

TIPPETT: And the word "stranger" in there, which is the complete alienation not only from others but from yourself.

BARROWS: Ah, from oneself, exactly. That's the worst of it.

TIPPETT: There's a paradox here that's running through all the conversations I'm having about this subject, and you bring it up again. Depression eventually can yield maturity and growth and a kind of spiritual insight—"a bigger soul" is the way some people might say it. But in the moment, in the depth of the experience, that kind of reflection is what is completely out of the question.

BARROWS: Yes, exactly.

TIPPETT: What does that mean?

BARROWS: All of the talk about, "Oh, well, this will be really good for your soul or your character, this will make a better person of you," feels like absolute rubbish when you're in the midst of the wretchedness of depression. But in a way—it almost feels physiological. If the soul were material, depression works on it the way you could work a piece of clay. It softens and it becomes more malleable. It becomes wider. It becomes able to take in more. But that's only afterward. In the fire, what you get is the fire.

This poem is called "Questo Muro." It is a phrase from a passage in Dante's *Purgatory*. Dante has been in the depths of depression, in the depths of the inferno, and he's now working his way out of it towards Beatrice, who you could call the soul or the *anima*. And he and Virgil are climbing the mountain, and all of a sudden they get to a wall of fire, and you can't go any farther unless you go through it. So this is my poem, and it really is a poem, I think, about finding the courage to persist, to go through that fire.

> You will come at a turning of the trail
> to a wall of flame
>
> After the hard climb & the exhausted dreaming

you will come to a place where he
with whom you have walked this far
will stop, will stand

beside you on the treacherous steep path
& stare as you shiver at the moving wall, the flame
that blocks your vision of what
comes after. And that one
who you thought would accompany you always,
who held your face
tenderly a little while in his hands—
who pressed the palms of his hands into drenched grass
& washed from your cheeks the soot, the tear-tracks—

he is telling you now
that all that stands between you
& everything you have known since the beginning

is this: this wall. Between yourself
& the beloved, between yourself & your joy,
the riverbank swaying with wildflowers, the shaft

of sunlight on the rock, the song.
Will you pass through it now, will you let it consume

whatever solidness this is
you call your life, & send
you out, a tremor of heat,

a radiance, a changed
flickering thing?

✻

Here in closing is another of Anita Barrows's poems, titled "Heart Work."

> Monday. Bronze sunlight
> on the worn gray rug
> in the dining room where Viva sits
> playing her recorder. Pain-ripened sunlight
>
> I nearly wrote, like the huge
> vine-ripened tomato
> my friend brought yesterday
> from her garden, to add to our salad:
> meaning what comes
>
> in its time to its own
> end, then breaks
> off easily, needing no more
> from summer.
>
> The notes
> of some medieval dance
> spill gracefully from the stream
> of Viva's breath. Something
> that had been stopped
>
> is beginning to move: a leaf
> driven against rock
> by a current
> frees itself, finds its way again
> through moving water. The angle of light
>
> is low, but still it fills
> this space we're in. What interrupts me

is sometimes an abundance. My sorrow too,
which grew large through summer
feels to me this morning

as though if I touched it
where the thick dark stem

is joined to the root, it would release itself
whole, it would be something I could use.

10

*

On the Complementary Nature of Science and Religion

"QUARKS AND CREATION"

I first heard Sir John Polkinghorne's voice on the BBC in the late 1980s, at a time when I lived in England. Late one night, he presented a riveting radio essay. It couldn't have lasted more than five or ten minutes, but it had a tremendous, lasting effect on me.

Polkinghorne spoke about reason and faith, science and prayer—subjects I was pondering deeply at that point, after a good decade in which I had dismissed religion and religious sentiments out of hand. He described connections between quantum physics and theology in inviting, commonsense terms. He applied chaos theory to make prayer sound intellectually intriguing. I was thrilled when I was able, in 2005, to talk with John Polkinghorne about the ideas he inspired in me fifteen years earlier and about many related questions I had accumulated since.

Just as I found myself speaking with him, of course, the centuries-old debate between science and religion—in particular the flashpoint of evolution versus creationism—was taking on renewed energy in American culture. And even as that debate receded from the limelight, figures like Richard Dawkins repopularized the thesis that scientific reason and religious faith are incompatible and at odds. But ironically, in this same historical moment, a lively, deepening international dialogue between scientists and religious thinkers was expanding across the rift that developed after Charles Darwin published *The Origin of Species* in 1859. John Polkinghorne has been a leading figure in that development.

Most striking, however, is how John Polkinghorne's perspective simply transcends the parameters and arguments that drive our cultural controversies.

Polkinghorne takes the Genesis stories, the biblical accounts of creation, seriously. But he points out that these are lyrical, theological writings. They were not composed as scientific texts. The early Christians, he says, knew this, and only in the later medieval and Reformation times did people begin to insist on literal interpretation. To read a work of poetry as a work of prose, he analogizes, is to miss the point.

Drawing on the best of his scientific and theological knowledge, Polkinghorne believes that God created this universe. But this was not a one-act invention of a clockwork world. God did something "more clever": he created a world with independence, a world able to make itself. Creation is an ongoing act, Polkinghorne believes, one in which the laws of nature make room for choice and action, both human and divine. He finds this idea beautifully affirmed by the best insights of chaos theory, which describes reality as an interplay between order and disorder, between random possibilities and patterned structure.

I'll let you read for yourself how he approaches mysteries like prayer, and the problem of suffering, in this frame of mind. But I will highlight two other evocative notions from our interview.

First, modern science increasingly suggests that contradictory explanations of reality can be simultaneously true. A scientific puzzle of whether light is a particle or a wave was resolved with the discovery that light has a dual nature as both a particle and a wave. And here's the key that made that discovery possible: how we ask the questions affects the answers we arrive at. Light appears as a wave if you ask it "a wavelike question" and it appears as a particle if you ask it "a particle-like question."

Second, there is the matter of quarks. Modern quantum physics has come to depend on quarks as a foundational element in understanding the way the world works. But in a very real sense, quarks are an article of faith. No scientist has actually seen one, nor do scientists necessarily ever expect to. They are believed to exist because the idea of quarks gives intelligibility to the whole of observable reality.

These scientific notions give me new, creative ways to imagine the credibility of religious modes of thought. They underscore John Polkinghorne's personable and passionate message that we need the insights of science and religion together to "interpret and understand the rich, varied, and surprising way the world actually is."

✳

Quarks and Creation

KRISTA TIPPETT, host
JOHN POLKINGHORNE, former Cambridge physicist,
Anglican priest

✳

*F*or twenty-five years, John Polkinghorne distinguished himself in the field of elementary particle physics as a professor at Cambridge. In 1974, he was named a Fellow of the Royal Society, the scientific academy to which Isaac Newton, Charles Darwin, and Stephen Hawking have also all been admitted. Then, at the age of forty-nine, Polkinghorne became a student again, this time of theology. He came to find scientific and religious questions to present a lively complement to each other, to be intellectual partners in discerning truth.

John Polkinghorne eventually returned to Cambridge to teach about the interface between science and religion. He's published many books and articles and emerged as one of the world's leading thinkers on the shared ground between the insights of quantum physics and religious mysteries. In Great Britain, he's chaired government initiatives to consider the ethical issues raised by cloning. In 1997, he was knighted by Queen Elizabeth.

And five years later, he won the Templeton Prize for progress in science and religion. Polkinghorne's vocabulary about God and science tends to stress qualities not often mentioned in science-religion debates—qualities such as beauty, subtlety, and surprise.

POLKINGHORNE: If working in science teaches you anything, it is that the physical world is surprising. And I was a quantum physicist, and the quantum world is totally different from the world of everyday. It's cloudy, it's fitful. You don't know where things are if you know what they're doing. If you know what they're doing, you don't know where they are. It's a complex world and quite different from what we expected. But it's an exciting world, because it turns out we can understand it. And when we do understand it, we have a deep intellectual satisfaction. Now, if the physical world surprises us and is different from everyday expectation—common sense, if you like—it wouldn't be very odd, really, if God also turned out to be rather surprising. Things that are on the surface easy to believe are not the whole story. There's a deeper, stranger, and more satisfying story to be found, both in science and in religion.

TIPPETT: I'd like to ask you about a few other words that you use, concepts where you bring together both theology and religion, and ask you flesh them out for me. Another one is beauty.

POLKINGHORNE: Well, beauty is a very interesting thing, and a form of beauty that is important to me is mathematical beauty. That's a rather austere form of aesthetic pleasure, but those of us who work in that area and speak that language can recognize it and agree about it. And we've found in theoretical physics that the fundamental laws of nature are always mathematically beautiful. In fact, if you've got some ugly equations, almost certainly you haven't got it right and you should think again. So beauty is the key to unlocking the secrets of the physical world.

TIPPETT: What are the qualities and properties—how do you describe what's beautiful about a mathematical equation?

POLKINGHORNE: It's very hard, of course, to describe any form of beauty. In some sense you have to perceive it. And it's more difficult with mathematics, because you have to be able to speak the language. It's a bit like saying, "This is a wonderful Icelandic poem," but if I don't understand Icelandic, I won't get the gist of it. Mathematical beauty is connected, first of all, with things being elegant and economic. You don't write a great sprawling equation that takes half a page to write down. It's very concise, just perhaps a line with only a few symbols in it. But it turns out that it's also very deep. This very simple-looking thing implies this, it implies that, all sorts of surprising and unexpected things. And if it's a successful part of mathematical physics, of course, it will imply all sorts of phenomena happening in the world. That's what we mean by mathematical beauty. It's very hard in everyday language to get a closer description of that. What is striking, I think, is that those of us who happen to speak that sort of language can agree about mathematical beauty. In fact, I suspect we agree rather more readily about mathematical beauty than, say, painters do about artistic beauty.

TIPPETT: You've also talked about how, in the same way that we take seriously the insights of science, we need to listen to the words of poets and to the insights of saints and mystics.

POLKINGHORNE: Absolutely. Yes. I think reality is very rich, many-layered. Science, in a sense, explores only one layer of the world. It treats the world as an object, something you can put to the test, pull apart and find out what it's made of. And, of course, that's a very interesting thing to do, and you learn some important things that way. But we know that there are whole realms of human experience where testing has to give way to trusting. That's true in human relationships. If I'm always setting little traps to see if you're my friend, I'll destroy the possibility of friendship between us.

And also where we have to treat things in their wholeness, in their totality. I mean, let's say a beautiful painting. A chemist could take that beautiful painting, could analyze every scrap of paint on the canvas, tell you what its chemical composition was, would incidentally destroy the painting by doing that, but would have missed the point of the painting, because that's something you can only encounter in its totality. So we need complementary ways of looking at the world.

TIPPETT: When you talk about moving from testing to trusting, scientists do that, too. Right? I mean, quarks have become an explanation, but isn't that something that scientists also take on faith, in a sense?

POLKINGHORNE: Quarks are in some sense unseen realities. Nobody has ever isolated a single quark in the lab. We believe in them not because we've, even with sophisticated instruments, seen them, so to speak. But because assuming that they're there makes sense of great swaths of physical experience. I was lucky enough to be a humble member of the particle physics community during the time all that was being worked out, and it was great fun to be, in a small way, part of it.

TIPPETT: I should ask you to explain quarks.

POLKINGHORNE: When I began many years ago as a graduate student working in science, we thought that matter, nuclear matter, was made up of protons and neutrons. And then, as we experimented and as we began to find out more and more about what was going on, it became more difficult to understand things in those terms. And it gradually dawned on people, it dawned on some very clever people, that maybe the protons and neutrons themselves were made up of something yet smaller, yet more basic, and these would have some quite surprising properties. For example, they would have fractional electric charge, which nobody has ever seen directly.

And then people began to see that, though they couldn't see these entities on their own, the way matter behaved—the way it was organized, the patterns of structure that it had, the way projectiles bounced off target particles—all that made sense if these unseen quarks were sitting there inside, never capable of being knocked out, but nevertheless real. In this indirect way, the unseen reality of quarks became an absolutely fundamental aspect of our understanding of the structure of matter. That remains the case. And I with all particle physicists believe, very fervently in a way, in the reality of quarks. But it's an unseen reality. It's the fact that they give intelligibility to the world that makes us believe that they're actually there.

TIPPETT: It's such a fanciful word, "quarks." How did they get named?

POLKINGHORNE: One of the people who made a great deal of these discoveries was an American theoretical physicist called Murray Gell-Mann, who is also a polymath sort of person; he's very interested in language. He had read James Joyce's *Finnegan's Wake*, and there's a line in there which says, "Three quarks for

Muster Mark!" Quarks come in threes, and so Murray picked that up and made it his.

TIPPETT: It's a literary word.

POLKINGHORNE: It's a learned literary joke, I think.

TIPPETT: I love that. There's something you wrote that I thought made sense in terms of this idea that you can be a scientist and a religious person and take seriously the insights of both and not necessarily find them to be in opposition. You've talked about how wave and particle theories can both be true—and about how Paul Dirac in 1920 at Cambridge suddenly made it clear how light could give a wavelike answer if you asked a wavelike question or a particle-like answer if you asked a particle-like question. Can you explain what he's describing and what that means to you?

POLKINGHORNE: It's a very striking example of how surprising the physical world is. People had been arguing about what light was like for a long time. Newton had some ideas about it. In the nineteenth century, people made some discoveries, both experimental and theoretical, that clearly showed that light behaves like waves. There are certain properties of waves, which showed up in an absolutely unquestionable way, and so the answer seemed to be settled. But right at the beginning of the twentieth century, through the ideas of Max Planck and also a young chap—an examiner in a patent office in Bern called Albert Einstein—people saw that light also had particle-like properties. That was a real crisis, you see, because a wave is a spread-out, flappy thing. A particle is a little bullet. So how could something be sometimes spread out and sometimes bulletlike? For about twenty-five years nobody knew. But the scientists just had to hold on to experience by the skin of their teeth even if they didn't quite know how to reconcile it.

And then the thing has a happy ending, I'm glad to say back in my old University of Cambridge, where Paul Dirac discovered something called quantum field theory. Now, a field is something that is spread out and it can be flappy. It certainly has wavelike properties. But when you bring in quantum theory, it makes things come in packets. That's the effect of quantum theory; it chops things up into little packets. Little packets look like little particles, so a quantum field has both these sorts of properties. And if you ask it a wavelike question, it gives you a wavelike answer. You ask it a particle-like question, it gives you a particle-like answer. And, you can't ask both questions at the same time, which saves you from having, you know, a contradiction.

TIPPETT: But you take both answers into account?

POLKINGHORNE: You take both answers into account. And the important thing I want to emphasize is that people had to cling on to taking both insights into account before they understood how they fitted together. We don't make progress by chopping experience down to a size that fits into our current theories. We have to allow the way the world is to modify our understanding of the world. And if you're a Christian theologian, and you're telling that sort of story that I've just told about light being both particle-like and wavelike, we know that the Christian story about Jesus Christ is that he is a human being but also, in some real sense, needs to be described in terms of divine language. It's the same sort of dilemma, if you like. And we're not quite so clever, theologically, at finding the precise answer to that. But, again, we don't make progress by denying our experience.

TIPPETT: Right. I was going to say that that model, the paradigm that you don't explain things and come to more wisdom by squeezing things into a theory—maybe scientists are more open to that way of moving through the world than religious traditions sometimes have been.

POLKINGHORNE: They may be. Actually, scientists don't find it easy to change their views either. People sometimes say scientists question everything all the time. Of course they don't. We would make no progress if we did. If you were an eternal skeptic, you'd never get anything done. So it's painful and difficult. But there are times scientists do allow experience to mold their thinking. And perhaps more slowly, I think, religious people do, too, but it's not quite so quick.

TIPPETT: So one of the ways religious people dealt with science for a long time—and I'm talking about the last couple of centuries—was what we've now called the "God of the gaps" idea. Describe your understanding of how that worked. People in the science/religion dialogue refer to that a lot, but I don't think laypeople have a memory of it.

POLKINGHORNE: When science first came into being in the seventeenth century and then in the eighteenth century became very successful through the discoveries of Newton and the aftermath of all those, some people began to say, "Okay, science can explain the solar system. It can explain everything." And the religious people tried to fight back by saying, "No, science can't explain everything. There are gaps in our knowledge, which only God can fill." For example, the human eye is a very complicated, a very beautiful optical system. How could that have come about other than being made, so to speak, directly by God? Of course, then Charles Darwin came along in the nineteenth century and showed how the eye could have evolved piece by piece, slowly and slowly, and drew the rug from beneath that argument.

People could see with hindsight that the God of the gaps type argument—the God who's stepped in to do the things that science couldn't currently explain—was in itself a theological mistake. If there is a God who is the God who is the Creator of the world, that God is the God of the whole show, not a cosmic stunt artist who does the difficult things, the obscure bits, and leaves nature to do the rest. It's back to this fundamental mistake of feeling that if nature does it, we don't need God. God is the God who ordains nature. God works through nature as much as through anything else. These days, in the science and religion community, the most contemptuous criticism you can make of somebody is to say, "I think your argument is a 'God of the gaps' type of argument." So we've learned something—something that is theologically helpful.

TIPPETT: There's a passage in Dietrich Bonhoeffer's *Letters and Papers from Prison* where he's reflecting on that in the mid-twentieth century. He writes that if God is consigned to the unknowable and we're learning more and more, then God is always being pushed further and further out of human experience.

POLKINGHORNE: That's right. The God of the gaps was a sort of Cheshire Cat deity, fading away with the advancement of knowledge. But actually, again we're back to this question of truth. If God is the god of truth, then the more truth we have, the greater understanding we have, the more we are actually learning about God.

✳

In John Polkinghorne's book *Quarks, Chaos & Christianity*, he writes, "We can take with absolute seriousness all that science can tell us and still believe that there is room left over for our action in the world and for God's action, too." He continues, "Of course, this does not mean that prayer is just filling in a series of blank cheques given us by Heavenly Father Christmas. Prayer is not magic. It is something much more personal, for it is an interaction between humanity and God."

✷

POLKINGHORNE: Of course, there are all sorts of different forms of prayer. There's worshipful prayer. I think a lot of scientists actually pray in that way without knowing that they're doing it, because one of the rewards for what is actually a laborious business doing scientific research is a sense of wonder when you see the beautiful structure of the world or the way things work. And though scientists don't use the word "wonder" when they write formal papers for learned journals, they use it quite a lot in their conversation. It is, as I say, the payoff for all the labor. That actually is a form of worship, whether the scientists know it or not. But I suppose the crunch question is, can a scientist ask God to do something? A petitionary prayer in that sense.

TIPPETT: Knowing what you know about the laws of nature and, in fact, respecting that it works and how it works?

POLKINGHORNE: That's right. Well, if the world were clockwork, then I suppose you'd have to hope that God had designed the clockwork and wound it up in such a way that things wouldn't turn out too badly. But twentieth-century science has seen the death of a merely mechanical and merely clockwork view of the world. It came first of all through quantum theory. At the sub-atomic level, quantum events are not precise and determinate. They have a certain randomness to them. They have a certain cloudiness to them. And we've learned, of course, from chaos theory, the "butterfly effect"—very small disturbances producing enormously big consequences—that even the everyday world described by the physics that would have been familiar to Newton isn't as clockwork as people thought it was.

So the world is certainly not merely mechanical. I think actually we always knew that, because we have always known that we are not mechanisms. We are not automata. We have the power to choose, to act in the world. It's a limited power. We can't fly, but we have the power of agency. And if we can act in the world, then there's no reason to think that God can't act in the world as well. So twentieth-century science has loosened up our view of the physical world, and it's a world in which we can conceive ourselves as the inhabitants and acting in it and helping to bring about the future. I believe also in God. So my answer will be that scientists can pray. Not, of course, as magic, but as cooperating with God, if you like, to bring about the best for the future.

TIPPETT: I told you this before we began. About fifteen years ago I first heard your voice on the BBC late one Saturday night. I was not a scientist asking that question. I was a person who had been completely political asking that question. And you, in five minutes, gave me a way to think about it. You were talking about how you understand how the world works. That things function in their essence and move forward all the time: we breathe, the grass grows. But there are also places of randomness and little openings in reality, and you also imagine that as relevant to the idea of prayer.

POLKINGHORNE: Yes. Again, the old eighteenth-century pic-ture was a clockwork world. And there are, certainly, clocks in the world. The sun is going to rise tomorrow. We can tell you the exact minute at which it's going to rise. But we've also learned that there are lots of clouds in the world. That's to say, it's a process whose outcome is not clear and certain and is not clear beforehand. So it's a sort of mixture of the two. And that has a consequence for prayer. There are some things that it isn't sen-sible to pray for. An early Christian thinker called Origen, who lived in Alexandria, which is jolly hot in the summer, said you shouldn't pray for the cool of spring in the heat of summer. The seasons are going to be there. And of course, theologically, we think that the regularity of the seasons reflects, if you like, the faithfulness of the Creator. But there are other aspects of the world which are cloudy, and I think those are the areas where there is, so to speak, room for maneuver. And I think it's through exploiting that room for maneuver that we act in the world and that God also acts in the world. So there are other things that we can pray for. Even the weather is not just clockwork. So, though it might cause a bit of a shiver to run down some people's spines, I think we can pray for rain if we're afflicted by a drought.

TIPPETT: Give me another example. Rain is one, but what would be another example of openings for human action?

POLKINGHORNE: Most of life, actually, is cloudy, and in these cloudy areas, things can, so to speak, go either way. Recovery from illness is one. Of course, there are clearly illnesses that are mortal illnesses. There is a clockwork side to illness, if you like. But we also know that illness is very much affected, prior to recovery, by people's personality and so on. We can pray that somebody may be strengthened or encouraged or given hope, and that may very well lead to a form of healing that might not have been possible without that. The point is, if God acts through these cloudy processes and we act through these cloudy processes, we can't take them apart and say, "Okay, I can see that God did that bit," because we just can't itemize them. We can't perceive it directly. But by faith we may have the intuition that God is indeed working in that sort of way. There is going to be an ambiguity in interpreting these things.

TIPPETT: So this is about ambiguity and variables that we may not be able to perceive at any given moment.

POLKINGHORNE: That's right. But life is like that. We can't have it cut and dry. That enables us to be what we are. There's a very interesting scientific insight which says that regions where real novelty occurs, where really new things happen that you haven't seen before, are always regions which are at the edge of chaos. They are regions where cloudiness and clearness, order and disorder, interlace each other. If you're too much on the orderly side of that borderline, everything is so rigid that nothing really new happens. You just get rearrangements. If you're too far on the haphazard side, nothing persists, everything just falls apart. It's in these ambiguous areas where order and disorder interlace, where really new things happen, where the action is, if you like. And I think that reflects itself both in the development of life and in many, many human decisions.

❋

John Polkinghorne's perspective on life and science largely transcends popularized arguments that set scientific reason and religion at odds. He has written, "Both science and religion are needed to interpret and understand the rich, varied, and surprising way the world actually is." He believes that God created the world, but he does not understand creation as God's one-time production of a ready-made world. And he points out that the biblical creation stories were not written as scientific textbooks. Here's a translation close to the original biblical Hebrew of the first few verses of Genesis 1: "At the beginning of God's creating of the heavens and the earth, when the earth was wild and waste, darkness over the face of the ocean, rushing spirit of God hovering over the face of the waters, God said, 'Let there be light.' And there was light."

I wondered how John Polkinghorne thinks about the intelligent design movement that has arisen in recent years as a response to the theory of evolution.

✳

POLKINGHORNE: I think that the intelligent design people ask some interesting questions. They look at the molecular level of life. They look at things like the blood clotting process. Or they look at the little things that make entities swim around, the cilia that make them go around. And they say, these are quite complicated systems even at this molecular level. They have several component parts to them, and we can't see how they would work unless you had all those parts in place. And so they ask, How could that have come about in an evolving way, bit by bit, piece by piece? In fact, that's how evolution seems to work. It seems to be, again, a sort of unfolding process, a bringing forth, if you like. So the intelligent design people ask some quite interesting questions, and the questions are, in principle, scientifically answerable. But I don't think we yet know the answers. So I'm very cautious about the line of argument they're trying to make.

TIPPETT: You also bring your theology and your science together interestingly in seeing that there are things going on in the world, including human beings' interaction with nature at any given time, that represent competing freedoms. I think that's a very interesting, complex idea.

POLKINGHORNE: I think we live in a world of true becoming. That's to say, I don't think that the future is fixed; I don't think God fixed it. I think God allows creatures to be themselves.

TIPPETT: Does God know the future?

POLKINGHORNE: If we live in a world of true becoming—so that we play our little parts in making the future, and God's providence also plays a part in making the future, and the laws of nature that God has ordained play a part in constraining the form of the future—then actually even God doesn't know the future. That's not an imperfection, because the future is not yet there to be known. Now, that's a very controversial view, and not everybody has agreed with me about that, but that's how it seems to me. And I think that there's been a very important development in theological thinking in the twentieth century. It's reflected in quite different theologians, but they have this thing in common. They see that the act of creation, the act of bringing into being a world in which creatures are allowed to be themselves, to make themselves, is an act of love. It is an act of divine self-limitation. The theologians like to call it *kenosis* from the Greek word. God is not the puppet master of the universe, pulling every string. God has taken, if you like, a risk. Creation is more like an improvisation than the performance of a fixed score that God wrote in eternity. And that sort of world of becoming involves God's accepting limitations, and, I believe, accepting limitations not knowing the future. That doesn't mean, of course, that God will be caught out by the future in the same way that you and I are. God can see how history is moving, so to speak, but God has to react to the way history moves. Now to me that makes quite a lot of sense about the world.

TIPPETT: That's a kind of theological way of describing evolution—this creation that creates itself.

POLKINGHORNE: Yes. Absolutely. Darwin published *The Origin of Species* in 1859, and people think that was a great parting of the ways between science and religion, a big clash—with all the scientists shouting, "Yes, yes, yes," and all the clergy shouting "No, no, no." And that they just went their separate ways. Quite untrue. A lot of scientists had doubts about Darwin, actually, for a while. And some religious people—from the start, an English clergyman called Charles Kingsley said that God could no doubt have snapped the divine fingers and brought into being a ready-made world. But that God had done something cleverer than that: God had made a world in which creatures could make themselves. So that's the picture in which God brings a universe into being. It has great potentialities, great possible fruitfulness, but creatures are allowed to explore and bring that fruitfulness to birth. That seems to me a very beautiful and fitting form of creation, a better world, so to speak, than a world which was ready-made. But it has a necessary cost. It has a shadow side.

TIPPETT: Right. That's what I wanted to ask you, the theodicy question: if terrible things happen, what does that say about the nature of God?

POLKINGHORNE: Absolutely. The greatest difficulty, religiously, is the way the world is. It is beautiful and it's fruitful, but it's also ugly and terrifying. Dreadful things happen in the world. And the problem of evil and suffering is a very great problem. This scientific insight helps us a little bit with that. If creatures are going to make themselves, to explore this potentiality, there will be blind alleys and ragged edges in that exploration. That's bound to happen. A very simple example is this: the engine that has driven the three-and-a-half-billion-year history of life on Earth has, of course, been genetic mutation. For two billion years or so, there were only bacteria. Then things complexified, because genes mutated and new possibilities came along. So that's been a tremendous fruitfulness. But if that's going to happen, it's inevitable that other cells will mutate and will become malignant. You can't have one without the other. So, though the fact that there is cancer in the world is obviously an anguishing fact about the world, it's not, so to speak, gratuitous. It's not something that a God who is a bit more competent or a bit more compassionate could easily have eliminated. It's the shadow side of a world allowed to make itself.

TIPPETT: What does that way of looking at the world say about something like an earthquake or tsunami?

POLKINGHORNE: Well, if God allows creatures to be, God will allow tectonic plates to be.

TIPPETT: So creatures are allowed to be fully themselves— -not just human beings, but every aspect of nature?

POLKINGHORNE: When I say creatures, I'm thinking of the whole created order, the different parts of it. For example, we believe that having tectonic plates is an important necessity for a planet that's going to have life. Because between the plates, new material wells up from inside and replenishes the surface of the earth. But, of course, if there are going to be tectonic plates, not only will that happen, but sometimes they will slip. And when they slip, that will create an earthquake or, if it's under the sea, will create a tsunami. Again, it's a hard answer. It's not a . . .

TIPPETT: It's not a compassionate answer.

POLKINGHORNE: I think it has an element of compassion in it. But it's not a sentimental answer, that's for sure. There was this tremendous earthquake in Lisbon in 1755 that killed fifty thousand people in one day. And a great Oxford theologian said, "It was God's will." I think the hard answer instead was that the elements of the earth clashed and behaved in accordance with their nature. They are allowed to be just as you and I are allowed to be. It's not an easy answer, but I think, actually, it is the true answer.

TIPPETT: This is something I've come to understand through your work, this idea that free will is built in and that it's a gift, essentially. Human beings experience it as a gift. We're not robots. But earthquakes will be earthquakes, or tectonic plates also have their essence of being. Right? That's what you're saying.

POLKINGHORNE: That's right. They have their essence of being. And that is respected.

TIPPETT: And these freedoms—and this is the essential nature that's given to every aspect of creation—can collide and cause effects which will be devastating for one side or the other.

POLKINGHORNE: Yes, I think that's right. I think that God does respect the integrity of creation. God is not a magician or an interferer. I'm sure God interacts with the history of the world, but not in a way that overrules it. I believe that God wills neither the act of a murderer nor the incidence of an earthquake, but allows both to happen in a world which is a creation given a degree of independence by its Creator.

TIPPETT: But again, if the possibility of suffering is built into the creation, what does that say about the nature of God? Doesn't it take us right back to the age-old question of theodicy: How could a good god have made a world in which there is so much innocent suffering?

POLKINGHORNE: I'd want to say three things. First of all, the sort of argument we've been having at the moment is an intellectual argument. It's mildly helpful, but it doesn't, of course, answer all the problems. The problems with evil and suffering are deep existential problems. Why is this happening to me? Or, why is this happening to somebody I love? Those are entirely legitimate questions to ask. There's a particular Christian insight that seems very, very important to me, indeed, in some sense, enables the possibility of Christian belief. That is that the Christian God is not simply a compassionate spectator, invulnerable up in heaven, looking down on this strange and suffering world. But that God has also been a fellow sufferer, a fellow participant in the agony of creation. The cross of Christ, understood from the point of view of Christian theology, is God living a human life and nailed to the cross in the darkness and in the paradox of the dereliction—"My God, my God, why have you forsaken me?"—of Calvary. So God knows human suffering and the suffering of creation from the inside and not simply from the outside.

Also, I don't want to play a sort of pie-in-the-sky type of answer to things, but I do believe that this life is not the only life we live. I do believe we have a destiny beyond death. And though that doesn't explain away the sufferings of this world, I think they would be even more bitter, really, if there were no such destiny to look forward to.

TIPPETT: That's an article of faith, really.

POLKINGHORNE: Of course, as a Christian, I believe that it's an article of faith that has been exemplified and guaranteed within history by the resurrection of Jesus Christ. But it's not something with which we have direct experience.

There's a very deep human intuition of hope. Peter Berger makes this point very beautifully in a little book of his called *A Rumor of Angels*. He takes everyday things and says, think about them for a minute. Where are they pointing you? They're deeper than you think. For example, a child wakes up in the middle of the night, scared by a dream or something like that. A parent goes to the child and says, "It's all right." And Berger says, what's going on there? Is that a loving lie? Cancer, concentration camps—the world is not exactly just all right. But nevertheless, he says that is a deep human intuition. The assurance that that's so is an important part of enabling that child to grow up into full humanity. So there is a deep-seated human intuition of hope, the strangeness and bitterness of the world notwithstanding. And I do take that very seriously.

TIPPETT: You take that seriously?

POLKINGHORNE: Yes, I do.

TIPPETT: As part of the evidence we have of the truth we're trying to get at?

POLKINGHORNE: I think Berger calls these things "signals of transcendence," hints that take us beyond the everyday level of things. And I take it seriously at that level, yes.

TIPPETT: When I was learning about evolution when I was a child, I was told that that the only real leap that you have to make—at least with the Genesis 1 story of "In the beginning," and its progression of life forms—for that not to contradict with what we know from science is to say that God's days are longer than our days, that there's a different sense of time.

POLKINGHORNE: Well, not quite. This is an extraordinary thing, Genesis 1. It's the more sophisticated of the two stories, of course, and things don't quite come in the right order. It's striking that it begins with energy for light: "Let there be light." It's striking that life starts in the waters and moves onto the land. But the sun and moon and stars only come on the fourth day. And of course, there wouldn't be any life without the stars, because they make the raw material for life. So that isn't right. And we in theology believe that one of the reasons the sun, moon, and stars come downstream, so to speak, is that the writer is wanting to say the sun and the moon aren't deities. They're not to be worshipped.

TIPPETT: Because that was the conflict of his day.

POLKINGHORNE: They are created just like everything else. And that shows us that what we're reading is a theologically oriented thing and not a scientifically oriented thing. When you read something and you want to read it respectfully, you have to figure out what it is you're reading. Is it poetry or is it prose? If you read poetry and think it's prose, you will make the most astonishing mistakes.

TIPPETT: And Genesis 1 is a poem, isn't it?

POLKINGHORNE: It's much more like a poem than like prose. And that, in a sense, is the sadness of the "creationist." These people who are really wanting to be respectful to scripture are, I think, ironically being disrespectful, because they're not using it in the right way.

TIPPETT: So God's days aren't just longer. But let me ask you this. If it takes fourteen billion years to get to where we got now, by your understanding of the best of science that's out there, what does that long amount of time, that patience, say? How does that inform your understanding of the nature of God?

POLKINGHORNE: Well, certainly God is not a god in a hurry. That's clear. God is patient and subtle. God works through process and not through magic, not through snapping the divine fingers. That's what we learn from seeing the history of creation as science has revealed it. And I think that tells us something about how God acts generally. When you think about it, if God really is a God whose nature is best described as being the God of love, then that is how love will work. Not by overwhelming force, but by persuasive process. Again, it's an example of how religious insights about the nature of God and the scientific insights about the process of the world seem to me actually to be very consonant with each other. You can't deduce one from the other, but you can see that they fit together in a way that makes sense. They don't seem to be at odds with each other, and I find that encouraging.

TIPPETT: Another interesting and hard question, at least on the surface. If tectonic plates, which will always eventually create earthquakes and tsunamis, act according to their nature, how does that reflect on the idea that there is also some kind of moral nature to the universe?

POLKINGHORNE: Well, I don't think moral principles apply directly to tectonic plates. They apply to people who are moral agents. You could say that, therefore, if there are moral questions, they're about the morality of God.

TIPPETT: Right. Is God moral if God created tectonic plates?

POLKINGHORNE: Back towards something I said before, if God is going to bring into being a world in which creatures are allowed to make themselves, God does that because that is a greater good than a ready-made world or a magic world in which fire never burns anyone when they put their hands into it— where deeds, in fact, never have consequences. If that's a better world, then even God can't create that world without it having its shadow side. It's very important to understand what we mean when we say God is "almighty." We mean not that God can do absolutely anything. But God can do what God wills in accordance with God's nature. I mean, the good God can't do evil deeds. The rational God can't decree that two plus two equals five. And if God is going to bring into being a world in which creatures make themselves, and God judges that to be a world of greater good than a ready-made world, then even God cannot make that world a world in which there isn't a costly side to things.

TIPPETT: Are there any cutting-edge developments in your field of particle physics or in the world of science that really challenge your faith or that pose questions that you're holding in tension?

POLKINGHORNE: I've never, in my exploration of these things, felt I reached a crisis situation in which I was faced with an either-or choice, either go with science or go with religion. There are, of course, puzzles all the time. One of the things that's happening in the science-and-religion world, to some extent, in the last few years, it that people are getting interested in questions of what the theologians call eschatology.

TIPPETT: Which is the end-times.

POLKINGHORNE: Trying to make sense of the notion of a destiny beyond death. Then you raise questions of, what's the human soul? I don't think it's a detachable, spiritual bit. I think it's the real me. The real me is certainly not just a matter of my body, because that's changing all the time. Through wear and tear, eating and drinking, the atoms change. But the pattern in which the atoms are formed, there, I think, is what the soul is. This is what Thomas Aquinas, the great theologian in the Middle Ages, would have thought, too.

TIPPETT: That would be the pattern of your personality and your effect.

POLKINGHORNE: It's an immensely rich pattern. It doesn't finish at my skin. It obviously involves my memories, my character, my personality. It also, I think, involves all the relationships that help constitute me.

TIPPETT: It takes on substance in the course of your life.

POLKINGHORNE: Exactly. And that's very complex, and obviously we're struggling to even say something about it. But that's what it is, I think. And I think God will not allow that pattern to be lost, and God will re-create that pattern in an act of resurrection. These are the sort of things that people are exploring at the moment. What's happening is that the science and theology conversation is getting more theological. Theology is being allowed to set more of the questions. For a long time, and quite rightly for a long time, science set the questions. Here's Big Bang cosmology. Here's biological evolution. What do you make of that? Theology would seek to respond, and I think it's been able to respond pretty well. But now theology's asking some of the questions. What's the human person? What could be the carrier of continuity between life in this world and the world to come? And that's a healthy development. You want the conversation to be very even-handed in that respect.

ABOUT THE INTERVIEWEES

✳

FREEMAN DYSON is a theoretical physicist and professor emeritus at the Institute for Advanced Study in Princeton. He has published numerous scientific papers and written many articles and books, including *Disturbing the Universe*.

PAUL DAVIES is a theoretical physicist, cosmologist, and astrobiologist. He is director of Beyond: Center for Fundamental Concepts in Science at Arizona State University. He has written widely about frontiers in modern physics and about Einstein's understanding of time, including *How to Build a Time Machine*.

SHERWIN NULAND is clinical professor of surgery at Yale University, where he also teaches bioethics and medical history. His books include *How We Die* and *How We Live* (originally published as *The Wisdom of the Body*).

MEHMET OZ is vice chair and professor of surgery at Columbia University. He directs the Cardiovascular Institute and Complementary Medicine Program at New York–Presbyterian Hospital.

Dr. Oz appears regularly on *The Oprah Winfrey Show* and in 2009 he began hosting a syndicated daily talk show, *The Dr. Oz Show*. His books include *Healing from the Heart* and *YOU: The Owner's Manual*.

JAMES MOORE has coauthored several books about Charles Darwin, including *Darwin: The Life of a Tormented Evolutionist*. He has degrees in science, divinity, and history, and has been researching and teaching Darwin for more than thirty years in Cambridge, England. He is presently a visiting scholar at Harvard University.

V. V. RAMAN is emeritus professor of physics and humanities at the Rochester Institute of Technology. He has authored many papers on the historical, social, and philosophical aspects of physics and science, and has devoted several years to the study and elucidation of Hindu culture and religion. His books include *Truth and Tension in Science and Religion*.

JANNA LEVIN is an assistant professor of astrophysics at Columbia University's Barnard College. She's the author of two books, including *A Madman Dreams of Turing Machines*.

MICHAEL MCCULLOUGH is a professor of psychology at the University of Miami in Florida, where he directs the Laboratory for Social and Clinical Psychology. He's the author of *Beyond Revenge: The Evolution of the Forgiveness Instinct*.

ESTHER STERNBERG is a rheumatologist, researcher, and author of *Healing Spaces: The Science of Place and Well-Being* and *The Balance Within: The Science Connecting Health and Emotions*. She is also chief of the Section on Neuroendocrine Immunology and Behavior at the National Institute of Mental Health.

ANDREW SOLOMON is the author of three books, including *The Noonday Demon: An Atlas of Depression*, which was nominated for a Pulitzer Prize and won the National Book Award in 2001. He is a Fellow of the New York Institute for the Humanities.

PARKER PALMER is an educator, author, and activist who focuses on issues in education, community, leadership, spirituality, and social change. His numerous books include *Let Your Life Speak: Listening for the Voice of Vocation*. He is the founder of the Center for Courage & Renewal.

ANITA BARROWS is a poet, clinical psychologist, and translator of *Rilke's Book of Hours: Love Poems to God*. She is a faculty member at the Wright Institute, a graduate school for clinical psychology.

JOHN POLKINGHORNE is a physicist, theologian, and author of numerous books on science and religion, including *Quarks, Chaos, & Christianity*. He is a Fellow of the Royal Society and founding president of the International Society for Science and Religion. He is former president of Queen's College and professor of mathematical physics at Cambridge University.

ACKNOWLEDGMENTS

✳

This book reflects years of work by many people. In addition to Kate Moos, Mitch Hanley, Colleen Scheck, and Trent Gilliss, to whom this book is dedicated, special thanks goes to Andrew Dayton, Stevie Beck, Bill Buzenberg, Sarah Lutman, Nancy Rosenbaum, as well as Shiraz Janjua, Rob McGinley Myers, and Jody Abramson. At American Public Media, Bill King, Judy McAlpine, Mary Nease, Tom Kigin, and Mitzi Gramling have graciously helped make my work and this book happen. The Anam Cara Writers' and Artists' Retreat in Ireland offered wondrous hospitality, space, and beauty for writing, as did the Arc Retreat Center in Stanchfield, Minnesota, and my favorite everyday place to think (and eat), the Birchwood Café.

I'm grateful for other indispensable colleagues and mentors, consummate professionals who have enlarged this project with their knowledge and expertise: Steven Schumeister, Leah Stevenson, Tony Bol, my wonderful editor and friend Carolyn Carlson at Viking Penguin, and Scott Moyers of the Wylie Agency, agent extraordinaire, who sparked the original idea for this book and inspired it to the end.

None of this would be possible, of course, without the generosity of time, intellect, and spirit of my conversation partners and of our listeners. They have made what seemed a counterintuitive idea to some—intelligent public conversation about religion, spiritual ethics, and meaning—a reality and a success. And I would not be able to sustain or fully enjoy all of this without the anchoring presence and love of my children, Aly and Sebastian, who are my greatest pride and delight.

AVAILABLE FROM PENGUIN

Speaking of Faith

Why Religion Matters—and How to Talk About It

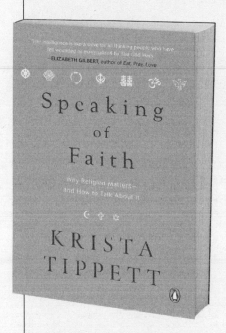

Krista Tippett draws on her own life story and her intimate conversations with both ordinary and famous figures, including Elie Wiesel, Karen Armstrong, and Thich Nhat Hanh, to explore complex subjects like science, love, virtue, and violence within the context of spirituality and everyday life.

"In a day where religion— or, rather, arguments over religion—divides us into ever more entrenched and frustrated camps, Krista Tippett is exactly the measured, balanced commentator we need."

—Elizabeth Gilbert, author of Eat, Pray, Love

ISBN 978-0-14-311318-8

PENGUIN
BOOKS